HISTORIC
WALLACE, IDAHO
AND MY UNFORESEEN TIES

HISTORIC
WALLACE, IDAHO
AND MY UNFORESEEN TIES

TONY BAMONTE

Tornado Creek Publications
Spokane, Washington

First edition published in 2017
by Tornado Creek Publications

Printed in the USA by
Walsworth of Marceline, Missouri

ISBN: 978-0-9821529-6-6
Library of Congress Control Number: 2017906107

Front cover and end sheet photos:
President Theodore Roosevelt during his 1903 visit to Wallace Idaho,
and the town of Wallace in 1887
Photos provided by Butch Jacobson and colored by Linda (Pachosa) Habbestad.

Tony and Suzanne Bamonte
P. O. Box 8625, Spokane, WA 99203 – 0625
Phone (509) 838-7114 Fax (509) 455-6798
Website: www.tornadocreekpublications.com
E-mail: tcpoffice@comcast.net

Dedications

To my father, Louis Bamonte, a miner, logger, and a good man

To Butch and Mary Jacobson for all their help in the research and photographs used in this project

To Colonel William Ross Wallace, the founder of Wallace, Idaho

Books by Tony and Suzanne Bamonte:

The Coeur d'Alene Gold Rush and Its Lasting Legacy
Spokane, Our Early History: Under All Is the Land
Life Behind the Badge: The Spokane Police Department's Founding Years, 1881-1903
Spokane's Legendary Davenport Hotel
Miss Spokane: Elegant Ambassadors and Their City
Spokane and the Inland Northwest: Historical Images
Manito Park: A Reflection of Spokane's Past
History of Newport, Washington
History of Pend Oreille County

Books by Tony Bamonte:

Sheriffs, 1911-1 989: A History of Murders in the Wilderness of Washington's Last County
History of Metaline Falls, Washington

About the Author

I was born in Wallace, Idaho, in 1942 and raised in Metaline Falls, Washington. My father was a logger and miner in both Shoshone County, Idaho, and Pend Oreille County, Washington. When I was six, my mother abandoned our family for another man, and they moved to Hawaii. It was then my father moved my brother, sister, and me to an area near Metaline Falls Washington. Although circumstances deprived my father from a formal education, he was smart, well-read, and, most important, an honest, hard-working man.

My father was exceptionally strict with us kids, but I grew up with great respect for him and the way he raised us. Among the many things I learned from him was the love of animals and a sincere caring about people less fortunate than we were. We always had pets in our family and, with the loss of each, I grew up more, learning that life is often short and temporary, for both man and animals.

During my early years, from the age of eight, I worked with my father in the woods, making fence post and cedar poles – later logging. Because of the G.I. Bill, I was able to obtain an education: A Master's Degree in Organizational Leadership from Gonzaga University and a Bachelor's Degree in Sociology from Whitworth College. I began a law-enforcement career as a military policeman in 1961. Following my discharge from the Army, where I served as a helicopter door gunner. I worked at the Pend Oreille Mines & Metals, and for Boundary Dam during its construction. In 1966, I joined the Spokane Police Department, where I worked for eight years, six as a motorcycle officer. Near the end of Expo '74, I resigned from the SPD and took a political appointment with the Sheriff's Office in Pend Oreille County. I was later elected and served as Pend Oreille County sheriff for over three consecutive terms, from 1977 until 1991. I have also been a relator, residential appraiser, and general contractor.

With my orphaned deer in 1985. This little guy would follow me everywhere I went when I was home. There are valid reasons you should not take in wild animals. However, it was either let him die or try to help. The day he died I lost a true friend.

Table of Contents

Foreword

Wallace is a mining town nestled in the mountains of northern Idaho. It is a gem that possesses more history than most towns of its size. The historic appeal the town now presents includes few signs of the dark side of its past. An untiring and determined researcher, Tony Bamonte does not gloss over the past, but rather exposes it for all to see.

I was very honored to be asked by Tony to write a foreword to this new book. Tony and I go back over 35 years in both a professional and social capacity. As a criminologist and professor at Eastern Washington University, I prevailed upon Tony (he was then the Pend Oreille, Washington County Sheriff) to lecture to different classes on a variety of criminal justice topics. His lectures were well received by my students. In addition to his lectures, he allowed my students to tour his jail and sheriff's operation.

During the course of our professional interactions we became personal friends. Many years ago he and his wife Suzanne began writing books through their company, Tornado Creek Publications. As I was also writing books through Chickadee Publishing, this common interest strengthened our friendship.

Tony is a dedicated and thorough researcher who gathers all of the facts before coming to any conclusions. He is never afraid to speak the truth, no matter the consequences. This has been a laudable and consistent characteristic over the years. It is clearly displayed in his numerous publications.

Having written a couple of books about prostitution in the town of Wallace, I was pleased to see another book about the town, especially one that discusses the town's founder, William Ross Wallace (Colonel, USA). Tony's book is not about prostitution per se, although he does cover it. The main emphasis is about the trials and tribulations that were foisted upon Colonel Wallace.

Through extensive research, Tony discovered where the Colonel was buried and the location of the tombstone that had been removed from his grave site. Over the years, Wallace was maligned about his military service (that he was not a real colonel). Tony's research set the record straight. Wallace served with honor during the civil war, even being wounded twice in combat.

Tony reveals how Col. Wallace was cheated out of ownership of the town he founded. This is a commentary on the greed many have when it comes to valuable items, in this case land. This injustice was perpetrated by local newspaper stories and the "legal" system.

Turning his attention to the town of Wallace, Tony discusses the twentieth-century raid by 150 FBI agents, the case of a serial killer, and prostitution. These topics are well-researched and fascinating reading. This book contains over 107 pages of historical photographs, many of which have never been published.

For Tony this book is important on many levels, as he was born in Wallace and has returned to his birthplace many times over the years. He discusses his family and their ties to Wallace, as well as the many dangers of hard rock mining; dangers that past and present residents are well aware of.

Tony Bamonte and Tornado Creek Publications have produced another thoughtful and enlightening review of Northwest history in this book about a small mining town that produced record amounts of minerals and prodigious amounts of history.

J. M. Moynahan
Professor Emeritus, Eastern Washington University

Acknowledgments

Butch and Mary Jacobson, are deserving of being his this book dedication. They not only opened their entire archive of photographic and printed materials, but also put us in touch with others who contributed in significant ways to the project. Whenever we needed something that was available only through local sources they wasted no time in responding.

Robert H. "Bob" Dunnsmore, a good historian from Wallace who provided me with the materials and photos that Butch Jacobson did not have. He was always there to be helpful.

Richard Magnuson, passed away in July 2016. His impressive knowledge of Wallace was invaluable. In terms of importance, it was surpassed only by the advice, support, encouragement, and humor he continued to offer along the way. Because Wallace was his lifelong home, he either knew personally or by reputation many people mentioned, and the events that occurred. He often had some insight from that incomparable vantage point. Hopefully, as we expressed it repeatedly, he knew how much we appreciated his involvement.

Fred Bardelli, another lifelong resident of the Coeur d'Alenes, studied every word of the manuscript from start to finish with a kind of dedication that would be any author's dream. His knowledge of the region's history and his attention to the finest details, in terms of both factual and grammatical presentation, was often mind boggling. We can only imagine the hours he spent working on the project, which resulted in immeasurable benefit. Fred's sister, Cleo Clizer, and her husband Gary Clizer, also contributed materials and photographs of their family. I am deeply appreciative of his family's continuous encouragement, support, and belief in this project.

John Amonson, a historian from Wallace, provided exceptional service in his role as director of the Wallace District Mining Museum (for nearly eighteen years), but as a friend, continued to be of help to us after retiring. Not only a source of information and resource materials, he was involved as a proofreader. John's valuable contribution to the preservation of the Coeur d'Alene Mining District's history has had a far-reaching effect. May this simple acknowledgment add to a collective "Thank you!"

Linda (Pachosa) Habbestad, for her great advice, artistic touches, and proofreading throughout the duration of this book.

We also wish to thank the following people and agencies or organizations (arranged alphabetically) for their various contributions of materials, photographs, or personal expertise (and, in some cases, proofreading):

Individuals: Patricia "Pat" Anderson, Rich Asher, Karen Curran, and Archie Hulsizer.

Agencies or organizations: Kootenai County Recorder's Office, Museum of North Idaho, Shoshone County Recorder's Office, Spokane Public Library – Northwest Room.

Photograph Credits

Linda (Pachosa) Habbestad for coloring book cover and end sheets
Butch Jacobson Collection
Robert L. Anderson Collection
Museum of North Idaho

Acknowledgments

The people pictured below assisted with or proofread this book project

John Amonson 1958 | Pat Anderson 2017 | Rich Asher 2017 | Jamie Baker 2017

Suzanne Bamonte 2015 | Fred Bardelli 1955 | Cleo (Bardelli) Clizer 1955 | Gary Clizer 1955

Dick Caron 2017 | Karen Curran 2017 | Linda E. Habbestad 1997 | Chuck King 2016

Preface

This book came about because of my personal ties to Wallace, Idaho – and because of the damnedest discovery about the town's founder, Colonel William Ross Wallace. I was born at the Providence Hospital on May 1, 1942. Although I only lived in the Coeur d'Alenes for a short time, I've had a lasting affection for the place of my birth, and for the early years my father spent working in the area's mines. When my family left Idaho, we moved to a similar logging/mining community. During my life, I have experienced many occasions that tied back to Wallace. The impetus to write the book followed a series of events that will result, on June 24, 2017, in placing Colonel Wallace's headstone in a position of honor in Wallace. After locating Wallace's burial site in Whittier, California, several years ago, I learned that his final resting place had been desecrated and made into a park. The headstone that had rested above Wallace's grave for 67 years had been removed and stacked in a vacant lot, along with other markers suffering the same fate. For 49 years, these headstones lay in a pile of other headstones. In 2017, through the efforts of friends Chuck King and Jamie Baker, the headstone was retrieved for the city of Wallace. Colonel Wallace left the area in 1889 after he lost the Wallace townsite through, what appears to be, some government ineptitude and people he had trusted and befriended turning against him. One of them was the owner of a newspaper, who was part of the group that was later granted the townsite patent. Newspapers have tremendous clout with politicians. One of the biggest mistakes people can make in life is to believe that the government acts in everybody's interest. And, as it pertained to many of the early newspapers – the news is what they say it is. (Read Chapter X for examples.)

As Baker and King began to plan a celebration around placing the headstone in a position of honor outside the Northern Pacific Depot Museum, Baker suggested I write a book about the town, as it had never been done. My wife, Suzanne, and I had just completed a 500-page book titled *The Coeur d'Alenes Gold Rush and Its Lasting Legacy*. The research for a comprehensive history book of this nature demands a great deal of time and effort. When we first began that book over 15 years ago, our plan was to cover the entire Coeur d'Alene Mining District (commonly known as "the Coeur d'Alenes"). During the process, we learned there was another book being written that was featuring many of the photos we intended to include. Consequently, we changed course and decided to focus on the gold rush, which initially brought the people to the Inland Northwest, and was the beginning of the Coeur d'Alene Mining District. The change was good, as nothing of major substance had ever been written about that area before – its history was little known. Our scope covered everything to do with the gold rush, including the towns and people, followed by whatever else of significance took place in the region encompassing the North Fork of the Coeur d'Alene River. In the research leading up to that book, we also unearthed considerable materials on Colonel Wallace.

As we researched, we soon learned that William Wallace was a decent, hardworking, and honest man, who had lost the Wallace town site as the result of deceptive people in high places using their connections to gain title to what he had worked hard to develop. In nearly every secondary source we read about the history of Wallace, when referring to Colonel Wallace, the writers often would add in parenthesis "not a real colonel." In truth and circumstance, William Wallace was a real colonel. He served in the Union Army for the duration of the Civil War, America's bloodiest conflict. The violence of numerous hand-to-hand, often bayonet-fixed, battles was a horrifying reality to the men who experienced it. Roughly two percent of the population, an estimated 620,000 men, lost their lives in the line of duty. Wallace was wounded in two separate battles and nearly lost his life in one of them. As it was, shrapnel imbedded in his head left him with frequent headaches up to the time of his death. Consequently, I feel Colonel Wallace is most deserving of whatever respect his namesake community would now give to him.

Wallace, Idaho, is a beautiful community, with a population of approximately 800. It is one of the most interesting and inviting communities in the Inland Northwest. It is located in a pleasant valley, surrounded by mountains with three creeks flowing into the South Fork of the Coeur d'Alene River – all within the city limits. The elevation is 2,728 feet above sea level. Wallace is also the county seat for Shoshone County, a county of approximately 13,000. It is in the heart of Coeur d'Alenes silver-mining district, which produced more silver than any other mining district in the United States. The entire downtown of Wallace is on the National Historic Register. For many years, the town was famous for having the only stoplight on 3,100-mile-long coast-to-coast Interstate 90, which ran through its business district. The site was first noticed by Colonel Wallace in 1883, when he was among the first to follow the flow of migration to a gold rush. Wallace came into being 22 years after the Mullan Road was completed. It was platted as a town site, originally called Placer Center, on May 1, 1884. Colonel Wallace was one of the first to recognize the silver and lead wealth within the vicinity of his future town site.

Wallace became the center of the mining industry. As such, numerous concentrator mills were built within the upper vicinity of the town – their waste flowing into the tributaries of the South Fork of the Coeur d'Alene River defiling the community and the environment. Hard-rock mining was among the most dangerous occupations. Prior to unions, a miner's life was not worth much to the mine owners. The fatality rate in the mines was high. In 1972, a fire broke out underground at the Sunshine Mine in Kellogg, Idaho, resulting in the deaths of 91 men from carbon monoxide poisoning. In his well-researched book *From Hell to Heaven*: *Death Related Mining Accidents in North Idaho*, the late Gene Hyde documented, with names and incidents, 3,238 deaths from mining accidents in the Coeur d'Alenes from 1887 to 2001.

In 1926, my father hitchhiked from New York City to San Diego to join the Marine Corps. While in the Marines, he became friends with another Marine who was from Kellogg. Because of the Depression, jobs were scarce. However, work could still be found in the mining district. He immediately went to work in the mines, and remained in the Coeur d'Alenes for 18 years. I was proud of my father. He was a great and humble man, and I was blessed to have had him for the short time I did. Writing this book gives me a chance to honor his memory. The cemeteries in the Coeur d'Alenes are full of men, like my father, whose lives were shortened by breathing the dust and toxic fumes within the mines. Men died due to the typical mine accidents, such as fires, cave-ins, gassings, drilling into misfired rounds, accidental discharges of dynamite, and falls down shafts, stopes or winzes. One of the greatest and ever-present threats to the miners was known as "miners' consumption" (medically known as silicosis). Over 40 percent of all miners suffered from miners' consumption. It killed many workers and often led to worse diseases, such as tuberculosis and pneumonia. Rock dust and unhealthy breathing conditions were the cause. The razor sharp dust cuts into a miner's lungs, causing a severe cough and eventual death. My father, who worked in the mines in the Coeur d'Alenes from 1930 to 1946, died from silicosis in 1963, at the at the age of 61. The last two years of his life were miserable for him and especially for his loved ones, who watched him suffer from this disease. I personally experienced the worry and heartache that was a result of my father's occupation.

Typical of other mining districts, prostitutes followed the miners to the areas they worked, and many settled in Wallace. The ratio of single men to women in the Coeur d'Alenes was profoundly unbalanced, creating a demand for prostitutes. The center for prostitution in the Inland Northwest was in Wallace, and lasted for over 100 years – in spite of weak laws against it. Prostitution was typically not an act of willfully violating the law. It was simply an act of supply and demand, and making a living in a small, but robust, community. It was women trying to make a living in a manner that worked for them, but it was fraught with many dangers – some of which will be addressed in this book. In Wallace, where prostitution was regulated, the majority of the local residents knew and accepted the prostitutes. They in turn gave back to the community, such as donating to high school sports activities and giving to causes that arose from various needs.

This book is the best way I know how to bring honor to Colonel Wallace and to promote the place of my birth. Hopefully, this will make up for the damnedest things that happened to W. R. Wallace.

Resurrecting the Reputation of William Ross Wallace, the Founder of Wallace, Idaho

The life story of William Ross Wallace, the founder of Wallace, Idaho, is shrouded in mystery and is marked by uninvited misfortune or downright bad luck. The facts of his life were elusive at best and recorded information frequently conflicting. Having an unusual name would have been helpful, because during our research, we found numerous early records for men named William Wallace.

When my wife, Suzanne, and I first began studying the history of the Coeur d'Alene Mining District nearly two decades ago, our impression of William Wallace, based on available secondary sources, was the one shared by the general population of the area – that he may have been a bit of a scoundrel and had lost his Wallace townsite by using Sioux scrip, subsequently ruled to be worthless. However, as we began to delve deeper, even engaging the services of a professional genealogist with remarkable skills, a different picture of Wallace began to emerge.

At the time of his death, he was lauded as a Civil War colonel who had served honorably under Major General William Rosecrans (one reference even stated that he served on Rosecrans's staff), despite frequent recent claims that "he was not a real colonel." He was described as a prominent mining mine, who was intelligent, hardworking, honorable, and generous to a fault. Although we set out to compile a "portrait" of the man who founded the town – platted and filed as Placer Center on May 1, 1884 – much of what we pursued continued to elude us.

This photo of William Ross Wallace appeared in the *Idaho State Tribune* on Wednesday, December 25, 1901, following his death on November 16, 1901. Note that he is referred to as colonel.

According to his obituaries, W. R. Wallace was born in Lexington, Kentucky, in February 1834, though the exact day is in conflict. Nothing could be found regarding his birth family or his early years, except that he had attended Amherst College. At age 27, he enlisted in the Union Army during the Civil War. According to various records, he served for the duration of the Civil War and was wounded at least twice. He subsequently engaged in mining, as did many other veterans, following the "next big bonanza." He was married three times, came to the Coeur d'Alenes during the gold rush, and soon platted the town of Wallace. After losing his townsite and being divorced from his second wife, he lived in Spokane for a short period, and eventually settled in Whittier, Los Angeles County, California, where he died at the Greenleaf Hotel on November 16, 1901. His wife, Annie E. Wallace, and stepdaughter, Annie G. Griffith, were with him at the time of death. He was survived by two other children: Oscar and Emma.

Although some of the above information is covered in more detail later in the book, this chapter is an attempt to repair the damage to his reputation, which has endured for over a century and a quarter, caused by some egregious attacks on his character and other undeserving indignities foisted upon this man in both life and death. As a result of my own personal experiences, some of which I have written about in this book, I have a great deal of empathy for what happened to William Wallace and some understanding, and certainly compassion, for how his townsite was taken from him.

A quick summary of what I most wish to address in this chapter is 1) the loss of his townsite, 2) the claim that he was not a "real" colonel, and 3) the desecration of his gravesite and headstone several decades after his death. Though minor in light of everything else, I feel compelled to mention one other indignity: Until a newspaper photograph of William Wallace surfaced through research for this book, Wallace residents had mistakenly accepted that a photo of his son, Oscar B. Wallace, was of William. It could almost be said of W. R. Wallace that if it weren't for bad luck, he would have had no luck at all.

Refuting the Issue of "Not a Real Colonel"

Every written thing that could be unearthed about William R. Wallace during his lifetime and at the time of his death referred to him as "Colonel." Yet, at some point in time and through some unknown source (at least to us), a tale emerged that he was not truly a colonel. Though we had no idea what this was based on, we felt that, if it were not true, it was a shameful attack on someone who had served in the most horrible war this nation has ever experienced – and it needed to be corrected. Simply by looking at some of the Civil War photos I have included in this section, it isn't difficult to imagine the daily horrors experienced by those who served either side. And if, in fact, Wallace was a true colonel, what would be the reason for denigrating his reputation in such a disrespectful way? Consequently, I was on a mission to find the truth.

Building coffins and burying the dead during the Civil War. *(Public domain)*

Some Brief Facts About the Civil War

The Civil War was an internal conflict fought in the United States from 1861 to 1865. The war began when the Confederates bombarded Union soldiers at Fort Sumter, South Carolina, on April 12, 1861. It ended in the spring of 1865, after General Robert E. Lee surrendered the last major Confederate army to Ulysses S. Grant on April 9, 1865, at Appomattox Courthouse, Virginia.

The estimated number of soldiers who fought for the Union were between 1.5 and 2.4 million, the majority of whom were volunteers. Estimates suggest five to six percent were drafted. The estimated total of Confederate soldiers range between 750,000 to one million. By 1845, the United States had about 7,300 men under arms to protect a nation of nearly 20 million and 1.8 million square miles of territory. By the end of the Civil War, somewhere between 2.75 and 3.5 million soldiers had fought in it.

The estimated number of combined North and South deaths that occurred from combat, accident, starvation, or disease during this bloody conflict range from 620,000 to 850,000. The number of wounded who at least survived long enough to muster out are estimated to be well over a million. Due to missing records, the number of casualties can only be estimated.

A medical examination was required to determine physical fitness for service. Every town had its own physician to do this work. The candidate for admission into the army had to remove his clothing, then was ordered to jump, bend over, kick, receive numerous thumps on the chest, back, and other areas of the body the physician thought necessary. The teeth and eyesight were also examined. If the candidate passed, he received a certificate of fitness. The recruit would then go to an assigned recruiting station, sign the roll of his new company or regiment, leave his description, including height, complexion, and occupation, and then accompany a guard to the examining surgeon, where he was again subjected to a critical examination as to fitness.

During the Civil War, weapons ranged from obsolete flintlocks to repeaters. Women took on new roles, including running farms and plantations, and spying; some disguised themselves as men and fought in battle. All of the nation's ethnic groups participated in the war. It was also a war that saw many new beginnings, including America's first income tax, the first battle between ironclad ships, the first extensive use of black soldiers and sailors in U.S. service, the first use of quinine to treat typhoid fever, and, **most significant, America's first military draft. (Called conscription)**

When the war began, President Lincoln and his staff were of the opinion that it would not continue beyond about three months. Consequently, a three-month enlistment was typical at the outset. They would soon realize otherwise. Barely had the "Three Months Men" reached the field before it became clear what a gigantic rebellion the United States had on its hands, and that a mistake had been made by not calling out a larger number of troops, and for a longer period of service. Because of that, on May 3, 1861, President Lincoln issued a call that United States volunteers and draftees would be obligated to serve three years.

In April of 1862, the Confederacy instituted the first draft in U.S. history. Clearly, the South, with a total population of nine million (including four million slaves), would have to muster all of its manpower to repel the North, which had a population of around 22 million. The Confederate draft exempted those who owned 20 slaves or more, which aroused resentment among the poor whites who constituted the vast majority of the army. Abraham Lincoln instituted a draft in the Northern states a year later, likewise calling on all able-bodied men, ages 18 to 35, to serve. There were exemptions in the North for draftees who could afford to pay a significant fee or provide a substitute.

Despite concerted efforts to locate W. R. Wallace's military records to determine whether or not he was a "real colonel," I could not prove with direct evidence that he was. However, despite never finding any references to his military service in secondary sources – aside from the allegation that he wasn't a colonel – there is solid documentation to support his service. There is also plenty of circumstantial evidence to support his rank as colonel, which is addressed later.

As another example of the "undeserved indignities," which no doubt befell many who served in the Civil War, some of Wallace's military records are missing. This is not surprising when the war is viewed from a broad perspective. Especially in the beginning, there were nearly insurmountable challenges. Shortly after President Lincoln's first inauguration, tensions that were already brewing escalated. After the April 12, 1861, attack on Fort Sumter, Lincoln declared a "state of insurrection." On April 15, having lost the fort to South Carolina, the President called for 75,000 volunteers to serve in the military for a three-month tour of duty. The response surpassed his request. Nobody expected a conflict of such overwhelming magnitude, and they were unprepared. The subsequent mobilization of men and resources proved to be the greatest undertaking in the history of U.S. government up until that point. It did not go smoothly.

Even today, with the help of computers and sophisticated databases, where records can be compiled in an orderly fashion, there is uncertainty as to the number of people who fought in the Civil War. Estimates of the total combined number of Union and Confederate soldiers range from 2.75 to 3.5 million. That is quite a disparity! The records that were kept during the war were, of course, handwritten and often prepared under abysmal conditions.

The U.S. National Park Service, assigned with the responsibility of maintaining those records, states on its website that, in the 1880s, the Department of the Army began to record the information on index cards. This resulted in 6.3 million General Index Cards, as they are termed. It takes no stretch of the imagination to conclude the risk of error in the monumental task of compiling that data, to say nothing of how many records may have been lost or destroyed during the war or that have become unreadable in the interim. Consequently, it was a disappointment, but not a surprise, when our friend, Karen Curran, a knowledgeable genealogist and skilled researcher, finally hit a dead end regarding proof of W. R. Wallace's rank as colonel. But it was not before exhausting all the records on two men named William Wallace who served under General Rosecrans, both of whom were believed to be colonels.

Military Records for William R. Wallace

Despite a lack of proof of the rank of colonel, the search was productive. On February 16, 2010, Curran received an e-mail response from Dennis Michael Edelin, Chief, Forms Reference Section, Textual Archives Division, which she forward to me. Regarding William R. Wallace's pension file, Edelin stated:

> *The records in our custody contain a compiled military service record (CMSR) for a* ***William R. Wallace, 1st Lt., Co. H 13th Ohio Volunteer Infantry****, an infantry regiment in the Union Army during the American Civil War (0.V.I.) (3 months), and it contains a Union Staff Officer file for a Colonel William Wallace, 1st Brigade, 2nd Division, 20th Army Corps. 15th Ohio Infantry.* [The latter proved to be another colonel by the name William Wallace.]

Curran was also able to obtain some copies of documents supporting Wallace's Civil War service, two of which are shown on the right. The education of William Wallace appears to have been of significance as it relates to his first enlistment, in 1861, into the Union Army. On his first day, though he enlisted for the three-month term as a private, Wallace was immediately promoted to a first lieutenant. His education was a likely factor, as that was among the acceptable criterion for rank and for being classified as an officer. In addition to his eduction, the promotion also suggests he comported himself well.

A piece of supporting evidence for his immediate promotion is shown on the facing page. The "Return" on the left, a record that was generated when he mustered out after his initial three-month tour

W | 13 (3 Months, 1861.) | Ohio.

Wm R. Wallace

1st Lt, Co. H, 13 Reg't Ohio Inf. (3 Mos., 1861).

Age 28 years.

Appears on Co. Muster-out Roll, dated

Columbus O Aug 14, 1861.

Muster-out to date Aug 14, 1861.

Last paid to No payment made, 186 .

Clothing account:

Last settled ...., 186 ; drawn since $......100

Due soldier $......100; due U. S. $......100

Am't for cloth'g in kind or money adv'd $......100

Due U. S. for arms, equipments, &c., $......100

Bounty paid $......100; due $...... 100

Remarks:

Mustered in Apl. 25. at Columbus.

Book mark:

Geo Dans

(361)

W | 13 | Ohio

William R. Wallace

1st Lieut, Co. H, 13th Reg't Ohio Vols. (3 mos)

NOTATION.

Book mark: R&P. 687396

Record and Pension Office,

WAR DEPARTMENT,

Washington, April 24th, 1902.

Under the provisions of the act of Congress, approved February 24, 1897, this officer is held and considered by this Department to have been mustered into the service of the United States in the grade of 1st Lieutenant, Co. H, 13th Reg't Ohio (3 mos) Vols. to take effect from April 25th, 1861, to fill an original vacancy.

W. M. Gibson

(442) Copyist.

In 2010, I asked an expert researcher and friend, Karen Curran, from Clarkston, Wash., to contact the National Archives of Military Service regarding Colonel William Ross Wallace's service records. The agency soon notified us they were unable to locate a complete record of his Civil War service. However, they sent the above copies of his "Returns." Both show Wallace's rank as First Lieutenant in 1861 during the three-month-enlistment term (April 24–August 14, 1861). The document on the right, from the Record and Pension Office, was created in 1902, when Wallace's widow, Annie Wallace, applied for a widow's pension. There were no similar records available showing when he enlisted for his three-year term, when he was promoted to colonel, or his second discharge. However, another record, spanning the years 1862 and 1863, documents the first time he was wounded (gunshot wound in his hand).

Major General William Starke Rosecrans (September 6, 1819–March 11, 1898) was an American inventor, coal-oil company executive, diplomat, politician, and U.S. Army officer. He gained fame for his role as a Union general during the American Civil War after having graduated from the U.S. Military Academy at West Point, New York, in 1842. Rosecrans served 12 years as an army officer and then resigned to become an architect and civil engineer in Ohio and Virginia. Returning to active service upon the outbreak of the Civil War, he served under Gen. George B. McClellan and Gen. John Pope, each of whom he succeeded when he moved east to larger commands. During 1862, Rosecrans led Union forces to victory in the battles of Iuka and Corinth, Mississippi, after which he moved on to Nashville, Tennessee, to take command of the Army of the Cumberland. He fought well at the intense but indecisive Battle of Stones River, or Murfreesboro (December 31, 1862–January 2, 1863). *(Public domain)*

REPORT OF MAJOR GENERAL ROSECRANS. 119

List of casualties in company B, 2d battalion 18th infantry, during the engagement before Murfreesboro', from December 31, 1862, to January 3, 1863.

Killed.—Lieutenant J. L. Hitchcock, Corporals John Limbaugh and Jacob R. Leibole, and Private Michael Gallivan. Total, 4.

Wounded.—Captain Charles E. Denison, solid shot in knee, leg since amputated; Sergeant William P. Leiboler, in leg, rifle ball. Privates Thomas P. Hunley, through arm and breast; Michael Maly, through both thighs; Roseline S. Conady, through right thigh; Patrick Mangan, through right arm; John Linament, flesh wound in leg; Edmund Coen, slight flesh wound; Martin H. V. Young, left breast and wrist, and William R. Wallace, slight wound, left hand. Total, 10.

H. G. RADCLIFF,
First Lieutenant 18th Infantry, Commanding Company B, 2d Battalion.

The above report, prepared by Major General Rosecrans, listed those in Company B of the 18th Infantry who were killed or wounded during the period from December 31, 1862, to January 3, 1863. Under the wounded is the name of William R. Wallace, who was, in fact, the founder of Wallace. Supporting documentation is printed on the facing page, a census listing William R. Wallace with "Gunshot wound in left hand." There was another Colonel William Wallace from the 15th who also served under Rosecrans, which complicated our search for the records on the founder of Wallace, Idaho.

of duty, indicates that he was a first lieutenant in Company H of the 13th Ohio Infantry. That record shows he mustered in on April 25, 1861 (a little over a week after the start of the war), but the record on the right from the Record and Pension Office of the War Department indicates he mustered in on April 24, 1861, and, effective April 25, 1861, he filled "an original vacancy," apparently as a first lieutenant.

After Lt. Wallace was officially discharged, as per his enlistment contract, because President Lincoln had since initiated the three-year term (see sidebar on page 3), Wallace again volunteered and served for the duration of the war. During that time, all the indicators lead to having attained the rank of colonel. According to his obituary, he served on the staff of General William Rosecrans, and later was in command of the Second Kentucky Cavalry.

During Wallace's service in the Civil War, he was wounded in action on at least two occasions. The document on this page, as well as the report by Major General Rosecrans, on the facing page, are corroborating evidence that he had reenlisted following the three-month term to serve a three-year term. According to Dennis Michael Edelin, Chief, Forms Reference Section, Textual Archives Division, Rosecrans's report is the one surviving military record of Wallace's second term.

PLACER COU

Auburn No. 3 Precinct, Cont

No.	NAME.	Business or Occupation.	Age	H'ght Ft.	H'ght In.	Complexion.	Color of Eyes.	Color of Hair.	Visible Marks or Scars if any, and their Locality.	Country of Nativity.	Local Residence Precinct.
115	Schwer, John Matthew	Carpenter	72	5	6	Medium	Hazel	Gray		Germany	Auburn No. 3.
116	Shepard, Will Abel	Printer	28	5	8	Medium	Hazel	Brown		Missouri	Auburn No. 3.
117	Skellinger, William Boyington	Carpenter	81	5	10	Medium	Hazel	Dark		New York	Auburn No. 3.
118	Slade, Charles Henry	Liveryman	37	6	1	Dark	Brown	Black		California	Auburn No. 3.
119	Smith, Ephraim "O"	Attorney-atLw	43	5	7	Medium	Hazel	Dark		Massachusetts	Auburn No. 3.
120	Smith, August	Miner	70	5	9	Fair	Blue	Gray		Finland	Auburn No. 3.
121	Smith, Edwin Guy	Miner	74	5	10	Light	Blue.	Gray		New York	Auburn No. 3.
122	Soper, Joseph Winslow	Invalid	69	5	8¼	Dark	Dk Blue	Gray	Right arm paralyzed; left hand crip.	Maine	Auburn No. 3
123	Spilman, James Franklin	Miner	69	5	10	Medium	Hazel	Gray		Kentucky	Auburn No. 3.
124	Stradling, William Charles	Brickmason	35	5	7¼	Light	Blue	Brown	Scar under left eye	England	Auburn No. 3.
125	Swensson, Samuel Edward	Clerk	56	5	10	Light	Blue	L't Brown.		Sweden	Auburn No. 3.
126	Taylor, Andrew Reuben	Carpenter	40	5	9½	Light	Blue	Light		Canada	Auburn No. 3.
27	Taylor, Joseph	Wheelwright	71	5	8¼	Light	Blue	Gray	Scar on left side of nose	N. Carolina	Auburn No. 3.
28	Teachout, John Edward	Laborer	23	5	7	Light	Blue	Brown		Michigan	Auburn No. 3.
29	Thompson, John	Miner	72	5	5½	Medium	Hazel	Dk Brown	Anchor tatooed on right hand	Finland	Auburn No. 3.
30	Tuttle, Fred Pierson	Attorneyat Lw	39	5	8½	Dark	Brown	Black		California	Auburn No. 3.
31	Tyler, Elias Mortimer	Blacksmith	65	6	1	Light	Blue	Dark		New York	Auburn No. 3.
32	Urich, Jacob Calvin	Farmer	47	5	8½	Dark	Brown	Dk Brown		Pennsylvania	Auburn No. 3.
33	Van Norden, Charles	Prof. Man	52	5	11	Dark	Hazel	Gray		New York	Auburn No. 3.
34	Van Vleet, Hazael	Tel. Operator	49	5	11¼	Dark	Brown	Brown		Ohio	Auburn No. 3.
35	Walker, Adolph Gustave	Blacksmith	28	6	¼	Medium	Lt. Brown	Brown		California	Auburn No. 3.
36	Wallace, Lee Ewing	Attorneyatlw	[illegible]	[illegible]	[illegible]	Blonde	Blue	Red		California	Auburn No. 3.
37	Wallace, William Ross	Miner	63	5	10½	Fair	Blue	Gray	Gunshot wound in left hand	Kentucky	Auburn No. 3.
38	Welsh, John Thomas	Hos. Steward	31	5	8	Dark	Brown	Black		California	Auburn No. 3
39	Walter, Jacob	Carpenter	64	5	6	Florid	Blue	Iron Gray	Left thumb crooked	Switzerland	Auburn No. 3.
40	Walter, Adam John	Watchman	33	5	11½	Medium	Brown	Brown		Ohio	Auburn No. 3.
41	Warner, John Erwin	Retired	79	5	8	Light	Gray	Gray		Pennsylvania	Auburn No. 3.
42	Waugh, Livingstone	Miner	39	5	7½	Medium	Gray	Dark		Scotland	Auburn No. 3.
43	Waugh, Robert	Hotelkeeper	42	5	10	Medium	Brown	Iron Gray	Scar bet. forefinger and thumb left hd	Scotland	Auburn No. 3.
44	Welsh, Nicholas	Retired	74	5	6	Light	Blue	Gray		Ireland	Auburn No. 3.
45	Wells, George Frederick	Miner	39	5	7½	Medium	Blue	Brown		New Hamps.	Auburn No. 3.
46	Will, Aloysius	Retired	80	5	9	Medium	Blue.	Gray		Pennsylvania.	Auburn No. 3.
47	Wiley, John Robert	Miner	64	6		Medium	Gray	Gray		Kentucky	Auburn No. 3.
48	Williams, Willis Richards	Retired	75	6		Medium	Gray	Gray	Paralysis of lower limbs	Maine	Auburn No. 3.
49	Williams, George Mariner	Cook	72	5	8	Light	Gray	Gray		Pennsylvania.	Auburn No. 3.
50	Witherspoon, Edgar Everett	Laborer	51	5	10½	Light	Blue	Lt. Brown.		Maine	Auburn No. 3.
51	Wooldridge, Edward	Nurseryman	40	5	11	Medium	Blue	Brown		Kentucky	Auburn No. 3.

In the 1896 Auburn No. 3 Precinct for Placer County, California, voter registration, William Ross Wallace is listed on line 37. Beside his name are the following: Business or Occupation: Miner; Age: 63; Height: 5 ft 10 1/2 inches tall; Complexion: Fair; Eyes: Blue; Color of hair: Grey; Visible Marks or Scars if any and their locality: Gunshot wound in left hand. This last entry confirms the injury reported by Major General Rosecrans on the facing page.

At the time of this census, William and Annie Wallace were living in Auburn, the county seat of Placer County, California. Auburn is known for its California gold rush history and is registered as a California Historical Landmark.

The foregoing report by General Rosecrans establishes that Wallace's first battle wound occurred sometime between December 31, 1862, and January 3, 1863. Unfortunately, the facts or conditions surrounding his second wound, which was to his head, are unknown. However, at the time of his death, his wife, Annie Wallace, requested an autopsy. It revealed shrapnel in his skull. According to Annie, he had suffered with severe headaches since the war. Her statement corroborates the occurrence of the second wound sustained in battle.

Circumstantial Evidence Regarding Wallace's Rank as a Colonel

Following his service in the Civil War, as a vocation, Wallace became involved in the mining industry in a number of states, including Minnesota, Texas, New Mexico, Arizona, Idaho, Washington, and in Canada. He consistently moved in circles with other high-ranking Civil War officers, which would have been unlikely had he, himself, not had a notable position of rank. Following are a few examples. To begin with, according to one of his obituaries (page 25), General U. S. Grant was a personal friend. He was also in a mining brokerage business with General A. P. Curry after he moved to Spokane in 1889. As the postal inspector at the time the town of Murray was founded in 1884, Curry gave the new post office his name, so for a short time, the town was simultaneously called Curry and Murrayville (soon shortened to Murray). In 1890, Wallace was appointed deputy attorney by Captain John Mullan, who was responsible for the construction of the Mullan Military Road and then attorney for the Coeur d'Alene Tribe, to participate in the ceding of the Coeur d'Alene Reservation lands to the government. Also note the connections in the story on page 26, "Colonel Wallace's Famous Dog." In addition to these select examples, the fact that he was consistently, and respectfully, referred to as colonel in written accounts (aside from the Wallace papers) should, in my opinion, silence the naysayers – and afford him his due respect for his years of military service.

These photos were probably taken within an hour after a battle on May 3, 1863, along the Sunken Road at Fredericksburg. The dead soldiers were members of Barksdale's Brigade. *(Public domain)*

Digging graves and burying the dead during the Civil War. Many of the victims of the war were left to decompose until they could be buried. Typically, the deceased were buried where they fell on the battlefield. Others were buried near the hospitals where they died. At most battlefields, the dead were later exhumed and moved to National or Confederate cemeteries, but because there were so many bodies, and because of the time and effort it took to disinter them, there are undoubtedly thousands, if not tens of thousands, of Civil War soldiers in unknown battlefield graves. An estimated 620,000 to 850,000 soldiers died from combat, accident, starvation, or disease during the Civil War. The smaller number came from an 1889 study of the war, performed by William F. Fox and Thomas Leonard Livermore, both of whom fought for the Union. The larger estimate is from a more recent study. *(Public domain)*

During the Civil War there were over a million horses killed. *(Courtesy Library of Congress)*

The 13-inch mortar, nicknamed "Dictator," which was mounted on a railroad flatcar, was one of many used during the Civil War. *(Public domain)*

Losing the Townsite

An unbiased free press is critical to the survival of our nation's democracy. From my own personal experiences as an elected official (a three-term sheriff of Pend Oreille County, Washington), I know first-hand how the media can wield its power for good, as well as for harm. The media often played important roles in my solving serious criminal cases, but it also played a major role in destroying my career in law enforcement. The media can truly make or break a politician, and sometimes it only takes one biased reporter to do serious damage. Consequently, most elected officials will bow to the will of the media. I discuss the power of the press in more detail in a later chapter, especially as it ties back to the town of Wallace, but I'm mentioning it here because, from my observation and assessment, I believe William Wallace's experience was similar to my own.

Wallace was eager to have a newspaper in his new town to promote and publicize it to the outside world, and he did what he could to support their efforts when John L. Dunn, of Portland, (and later his brother, Alfred J. Dunn) came looking to start a paper in the spring of 1887. Wallace provided a building, rent free for six months, as well as a commitment, along with a couple of others, to assist in widespread distribution of the paper. He also committed to regular paid advertisements. It was a two-way beneficial arrangement, with a signed six-month agreement, but as a result, the town's first newspaper, the *Wallace Free Press,* was successful. However, when the issue arose regarding the validity of the townsite patent, resulting from the use of Sioux half-breed scrip as payment, the tide began to turn against Wallace.

In *Coeur d'Alene Diary: The First Ten Years of Hardrock Mining in North Idaho* (Metropolitan Press, 1968), which is largely a compilation of local newspaper articles from those first ten years, Richard Magnuson wrote:

> *April* [1887] *brought interesting litigation to the district. In District Court, the Wallace townsite case came up for a hearing and* ***Lawyer Stoll asked that Al Dunn of the*** **Wallace Free Press** ***be excluded from the courtroom because his articles in the paper were adverse to his client's*** [***William Wallace***] ***interest****. The judge denied this request. Not content with this ruling Stoll subpoenaed Dunn as a witness and then Stoll moved to exclude all witnesses from the courtroom – and the judge was obligated to grant this motion.*

During the early years of the Coeur d'Alene Mining District's development, the Dunns had numerous other businesses in addition to the *Wallace Free Press*, including the *Coeur d'Alene Miner*, founded in 1890. With an understanding of the power of the press, they often used their newspapers for personal gain. It is my belief, based on various documents, that the Dunn brothers were responsible for vilifying Colonel Wallace and placing his townsite in a position of vulnerability by influencing both the public and elected officials. I found it rather suspicious that, after Wallace lost his townsite, John Dunn was among those to whom a patent was issued and then became chairman of the town's trustees (later the role of the mayor).

Because of the complexity of this issue, Chapter IV is devoted to the founding of Wallace and William Wallace's subsequent loss of the townsite.

Stranger Than Fiction: William Wallace's Gravesite Desecrated

In his final days of life, for health reasons, Wallace moved from Arizona to Whittier, California, where he died on November 16, 1901, at the Greenleaf Hotel. Accounts at the time of his death state he was buried at Mount Olive Cemetery, which is directly across the street from the Broadway Cemetery. All burials at the two cemeteries took place between 1888 and 1958. Due to lack of upkeep and use, the City of Whittier subsequently had the Mount Olive and Broadway cemeteries declared a nuisance.

In 1967, a citizens' cemetery committee began contacting relatives of persons interred in both the cemeteries to find out if anyone wished to have a relative's body moved, or to keep their headstone. Only seven people responded and had their relatives removed. In 1968, both cemeteries were cleared of

Pio Pico Park, where they were stored behind the old Pio Pico Mansion.

Following that, both former cemeteries were covered with topsoil and the land was turned into a public park, now called Founders Memorial Park. In 2001, the headstones were again moved, this time to the Whittier Museum, where they were stacked one on top of the other. There were typically five headstones to a stack, spaced approximately two feet apart, covering a large area of ground.

In 2010, I contacted the director of the Whittier Museum in an attempt to procure Wallace's headstone for the city of Wallace. She informed me how the headstones were stacked and that they were too heavy to move in order to look for any information on them. However, they did a search to see if they could find Wallace's headstone, with no luck. I then requested that, if they ever moved them and found it, to contact me. They stated they would, but I never heard back.

Chuck King's Involvement

In 2016, I passed this story to Chuck King, an avid history buff from Spokane Valley, Washington. Chuck was born in Spokane, the son of a former Spokane policeman (who was a friend of mine). He attended Whitman Grade School, after which the family moved to Spokane Valley, where he graduated from West Valley High School in 1976. At a young age, Chuck began collecting things such as coins and stamps – even rocks and shells. His family vacations always included historic stops along the way. As is the case with many people, as he matured, he became even more interested in the history around him. He then started collecting almost anything of a historical nature from around the area. His passion is researching local history and preserving the past. When he read about a proposed Spokane Valley museum, he became a founding member of the group that started the now-thriving museum.

Within a short time of hearing my story, Chuck located William Ross Wallace's headstone. He learned that, in 2015, officials at the Whittier Museum deemed the stacks of headstones to be a safety hazard and gave them away. Dale Bybee, a California collector extraordinaire, transported four semi-truckloads of grave markers to his Acton, California, estate, where he set up a "memorial cemetery" at the intersection of Sierra Highway and Red Rover Mine Road in California. The "cemetery" was a ruse to thwart the State of California's plan of building a high-speed rail line across his property.

After obtaining the information from the Whittier Museum, Chuck contacted Dale Bybee and learned he was, in fact, the new owner of William Ross Wallace's headstone. Bybee was pleased to be contacted about it and wanted to "make sure it was forwarded to the city of Wallace where it would receive the preservation effort it deserves." Although the memorial cemetery ruse failed when CalTrans discovered the "sacred" grounds were less than a year old and had markers but no bodies, it was a far more respectful sanctuary than being stacked in a lot next to the Whittier Museum.

Following his contact with Bybee, King engaged Jamie Baker, Wallace restaurant owner and hotelier, in an effort to obtain the headstone for the city of Wallace. The story of how that mission was successfully accomplished was written up by popular *Spokesman-Review* journalist Doug Clark and printed in the March 28, 2017, issue of the newspaper. It is reprinted in this book by permission.

Jamie and Barbara Baker

How many people would take the time to drive from Wallace, Idaho to Whittier, California, to pick up a headstone? Jamie and Barbara Baker were both game for that project. It took them nearly a week for the trip, and because the headstone weighed almost a ton, they borrowed a truck from an Osburn resident, Forest Van Dorn, to haul it to Wallace.

The Bakers own businesses and property in Wallace, including the Red Light Garage Restaurant, the Hercules Inn, and the former Al and Mae Hutton home. (The Huttons were remarkable people, who are written about in some detail later in the book.) The Bakers are history buffs who love the appeal of Wallace, Idaho, and are always eager to promote it.

CLARKSVILLE

HEADSTONE'S NEW HOME

KATHY PLONKA/THE SPOKESMAN-REVIEW

Jamie Baker talks March 21 about how he recently rescued the tombstone of Col. William Wallace, the founder of Wallace.

DOUG CLARK

Staff columnist

Wallace founder's marker saved from obscurity

With all due respect to Thomas Wolfe, it appears you can go home again.

Sort of.

Consider the strange case of Col. William Ross Wallace, who has returned to the North Idaho town he founded – in the form of his tombstone.

"William R. Wallace," reads the carved gray granite marker under a crest. "Died Nov. 16, 1901. Aged 67 years. A native of Kentucky."

No doubt about it, observed Jamie Baker, "Wallace should be in Wallace. He was the very

See **CLARKSVILLE, 7**

THE SPOKESMAN-REVIEW   MARCH 28, 2017 • TUESDAY • NEWS 7

FROM THE FRONT PAGE / BUSINESS

CLARKSVILLE

Continued from 1

first to come and try to get rich up here. Wallace was the first to do it."

The decorated Civil War veteran also left Wallace with some pretty hard feelings, but I'm getting ahead of myself.

Baker is a longtime friend and fellow Ferris High School bandmate of mine. Go Saxons!

He also operates the Red Light Garage, a landmark Wallace restaurant, with his wife, Barbara. A lover of Silver Valley lore, Baker is one of the key players in today's rather offbeat graveyard docudrama.

Last month, Baker embarked on an odyssey to Southern California, where he retrieved the 1-ton slab from a collector who had saved it from an uncertain fate.

For the time being, the Wallace headstone was laid rather unceremoniously to rest under a rumpled tarpaulin, just across the street from the Red Light.

But save the date, June 24.

On that day, Wallace's tombstone will be installed permanently near the entrance of the Northern Pacific Depot Railroad Museum as part of a major festival saluting the town's namesake.

"Shine or rain, we plan to have the biggest parade since Teddy Roosevelt came here in 1903," said Baker, who already has lined up Civil War re-enactors, a motorcycle club, high school bands, military and police escorts, civic groups, and even Abe Lincoln and Mary Todd impersonators. (Anyone wanting to join the fun should call Baker at (509) 435-3005 or the Wallace Chamber of Commerce at (208) 753-7151.)

This is sure to be a blowout. Even more so when you understand all the luck and circumstances that had to fit perfectly together in order to get the Wallace marker here.

The tale begins with Tony Bamonte, a Spokane writer of local history books and Wallace native.

In 2010, Bamonte and his wife, Suzanne, were researching material for a book on the "1883 gold rush to the Coeur d'Alenes" when they stumbled upon the unhappy story of what happened to Wallace's final resting place.

After his death in 1901, Wallace was buried in Whittier, California, and there he should have remained in perpetuity.

That lasted until progress got in the way in 1967. According to Bamonte, the city of Whittier declared two aged cemeteries as nuisances "due to lack of upkeep and use."

Wallace had been buried in one of these cemeteries, although there's some debate over exactly which one.

The important point is that all the vintage headstones, some 2,300 of them, were removed and stacked "in vacant lots and on state and city property," Bamonte said.

"Sadly, the gravestones were now separated from the people they were intended to honor and represent."

Among the dead was you-know-who.

Flash forward to 2016. Bamonte relayed the story about Wallace to Chuck King, a friend and fellow history buff.

The Spokane man began to hatch an idea. Wouldn't it be cool to try to track down the tombstone and get it to Wallace?

KATHY PLONKA/THE SPOKESMAN-REVIEW

The tombstone of Col. William Wallace, the founder of Wallace, sits across the street from the Red Light Garage in Wallace on March 21.

After some calls, King discovered that four semiloads of headstones had been hauled away by Dale Bybee, an antique collector from nearby Acton, California.

The electrical construction worker then arranged the tombstones on his property to look just like a real graveyard. Controversy erupted when Bybee was accused of building a "fake cemetery" as a ploy to stop a proposed high-speed rail project across his acreage.

Bybee doesn't deny it. He added, however, that he did what he did to "draw attention to the state trying to take my property."

He also pointed out that "most people who are collectors really like history."

As luck would have it, King finally contacted Bybee and discovered that the Acton man did, indeed, have Wallace's headstone.

It took a few more phone calls. But Bybee eventually saw the value in giving the marker to Wallace.

"You can't put a price on that stone," he said. "It's cool to be part of history instead of being on the sidelines."

And so the Bakers made their plans for a lengthy road trip. They drove to Mesa, Arizona, where Silver Valley businessman Forest Van Dorn spends his winters.

Barbara stayed behind while Baker and Van Dorn, who operates F&H Mine Supply, drove Van Dorn's Dodge pickup to Acton.

The tombstone was loaded into the truck with a forklift. Back in Mesa, Van Dorn let Baker drive his rig to Wallace with Barbara following.

Baker said he carried a letter of explanation about the unusual cargo in case he got stopped.

Some 3,400 miles later, the Bakers arrived back home. "Perfect weather all the way down and back," said Baker. "It was almost like Wallace was looking out for us."

It was a much better return for the town patriarch than the way he left.

Wallace, explained Bamonte, had come into North Idaho's "bustling 1883 gold rush area." He founded Placer Center, which was named Wallace in his honor in 1885.

He worked hard developing his town until 1889, said Bamonte, when he was swindled out of his property.

"Land-claim jumping was a common occurrence at that time and was taking place on a regular basis throughout the Coeur d'Alenes," he said.

About the same time Wallace's marriage to his second wife, Lucy, blew up.

There's no way to know, but it's a good guess that Wallace exited his town with a lot of hard and bitter feelings.

"His life was just crap," Bamonte said. "He was screwed over. He was maligned. But he was a hardworking and decent person."

And now Wallace, who was wounded twice while fighting for the Union during the Civil War, is getting the hero's welcome that he deserves.

"I think Wallace would be glad to be coming back in a different light," added Bamonte.

"I think it's great. There's a different attitude for him now. Circumstances are different and the people want him back."

CONTACT THE WRITER:
(509) 459-5432
dougc@spokesman.com

This article, which appeared on the front page of the *Spokesman-Review* on Tuesday, March 28, 2017, explains the efforts Jamie and Barbara Baker went through to relocate William Ross Wallace's headstone to the town of Wallace. It also details the plans he has coordinated with the Wallace Chamber of Commerce to set the headstone in a permanent place of honor out of respect for Colonel Wallace, as the towns's founder, and for his service in the Civil War.

At right, Jamie Baker's father, Jim Baker (standing at center) and his orchestra were the featured group in the Early Birds' Club at the Davenport Hotel from 1955 to 1968, where they broadcast live over the air. Jim played saxophone, clarinet, and flute. In addition to performing six nights a week, he also taught school. He played for the Friday afternoon Tea Dances from the late 1970s through the 1990s, and for various other special occasions, including New Year's Eve 1998, at age 84. During Baker's career, which spanned from age 16 to his death in 1999, he wrote 4,000 musical arrangements. Jamie Baker, also a talented musician, is second from left in the above photo, which was taken in 1974. *(Courtesy Jamie Baker.)*

The Ending of the Mount Olive and Broadway Cemeteries in Whittier, California

In the early days of Whittier, the two main cemeteries were Broadway and Mount Olive, located next to each other at Broadway and Citrus Avenue. This photo shows an early burial at Mount Olive, circa 1900. Over the years, the cemeteries became neglected and were finally abandoned. In 1968, all the headstones were removed from both cemeteries, and topsoil was spread over the area, to create Founders Memorial Park. A few of the bodies were moved to other cemeteries, but most still remain buried and unmarked in the public park. *(Public domain)*

Federal Law Prohibits Desecration of Military Graves

Although the following excerpt from the MILITARY AND VETERANS CODE SECTION 960-962 clearly prohibits the desecration of the headstones and grave sites of military veterans, that is exactly what happened to Colonel Wallace and many fellow military veterans.

ARTICLE 3. Care of Veterans' Graves [960 - 962] (Article 3 enacted by Stats. 1935, Ch. 389.

960. Whenever in any cemetery or place of burial of human remains, which is established or organized under the authority of the board of supervisors of any county or the governing body of any city, there is any known grave of a former soldier, sailor, or marine of the United States who was not dishonorably discharged from the service, the officers who manage such cemetery or place of burial shall keep such grave properly marked and identified, and free from weeds and rubbish, and keep in decent order and repair and free from defacement, injury, and unlawful markings any tomb, monument, gravestone, wall, or other appurtenance to such grave. (Enacted by Stats. 1935, Ch. 389.)

960.5. Whenever in any cemetery or place of burial of human remains there is any known grave of a former soldier, sailor, or marine of the United States who was not dishonorably discharged from the service, the board of supervisors of any county as to territory, whether incorporated or not, within it, and the governing body of the city as to territory within it, with the consent of the officers who manage such cemetery or place of burial, if any may keep such grave properly marked and identified, and free from weeds and rubbish, and keep in decent order and repair and free from defacement, injury, and unlawful markings any tomb, monument, gravestone, wall, or other appurtenance to such grave.

961. Any fraternal or benevolent organization which maintains a plot in a place of burial mentioned in section 960, which is devoted exclusively to the burial of soldiers, sailors, or marines of the United States, may apply under this article to the board of supervisors of the county in which the plot is maintained. Upon a showing of need, the board may keep the plot free from weeds and rubbish, and keep in decent order and repair and free from defacement and injury any tomb, monument, gravestone, wall, or other appurtenance to the graves in the plot.

962. The officers who are charged by law with raising money by taxation for maintaining any such cemetery or place of burial shall fix the tax levy at an amount sufficient to comply with the requirements of this article.

A small portion of Mount Olive Cemetery where William Ross Wallace was buried. This photo was taken before all the headstones were removed and the cemetery became part of a public park, named Founders Memorial Park. On August 2, 2009, a list was compiled of all the people buried in both Mount Olive and the Broadway cemeteries, totalling 2,315 individuals, many of whom were Civil War veterans. When the City of Whittier removed the headstones in 1968, there was no visible signage indicating who was buried there. That list has since been placed at the entrance to the park. *(Public domain)*

Stacks of headstones from behind the Whittier Museum in California. William Ross Wallace's headstone would have been among these. *(Public domain)*

The above headstone of William Whitchel, a Civil War veteran, was originally located in the Mount Olive Cemetery. Etched above his name (see detail upper right) are a cannon, cannon balls, and tents, which depict the fact that he was inarguably a Civil War veteran. There were numerous veterans buried in these two cemeteries. *(All photos on this page courtesy Dale Bybee)*

These two marker monuments are located at the entryway to Founders Memorial Park, formerly Broadway and Mount Olive cemeteries. A list of the people still buried there is etched on these monuments. There were originally 2,315 bodies buried in the cemetery. Seven bodies were removed and relocated by family members, leaving 2,308 still resting under the new park grounds. Among them, 12 were listed on the "official Whittier Cemetery list" with no information regarding which cemetery they were buried in. The block, lot, and grave numbers are unknown. William Ross Wallace was among the unfortunate few. Although there were definitive records of William Ross Wallace's burial at Mount Olive Cemetery (including one of his obituaries), because the records have no information, cemetery officials elected to put his name on the Broadway Cemetery marker.

Whittier, California, circa 1900. *(Public domain)*

Cemetery Information

LastName	FirstName	Age	Birth	Death	Block	Lot	Grave	Cemetery Name	Grave Marker	
Viramontes	Infant	Still		1919	8	2	3	Mount Olive	–	
Vistal	Frances L.		1878	1906	2	10	1	Mount Olive	–	
Viverez	Infant	Still		1933	7	18	2	Mount Olive	–	
Vore	William Townsend	67	1855	1923	7	25	8	Mount Olive	–	
Waddell	Frederick G.	26	1893	1919	10	23	8	Mount Olive	–	
Waddell	Kathrene Mary	22	1899	1921	10	23	7	Mount Olive	–	
Waddington	Frank	74		1916	6	21	7	Mount Olive	–	
Waer	Jean Mary	4	1926	1931	10	0		Mount Olive	–	
Waer	Emeretta A.	65	1852	1918	8	36	4	Mount Olive	35x36x12 (as of the 1990 list)	
Waer	James Nelson		1862	1926	8	36	6	Mount Olive	–	
Waer	James E.	10 m		1917	7	29	7	Mount Olive	–	
Waer	Mary Jane	71		1915	8	36	1	Mount Olive	piece (as of the 1990 list)	
Waer	Robert		1841		8	36	2	Mount Olive	35x36x12 (as of the 1990 list)	
Waer	William Thomas			1925	8	36	5	Mount Olive	–	
Wafford	Chas.	14	1877	1892	I	10	5	Broadway	–	
Wakefield	Infant	Still		1896	L	6	4	Broadway	–	
Waldron	Katherine E	65		1911	M	1	2	Broadway	–	
Walker	William H.C.	84	1836	1921	6	20	4	Mount Olive	–	
Walker	Ivey E.	29	1887	1916	3	12	2	Mount Olive	–	
Walker	Rebecca	84	1834	1918	6	20	2	Mount Olive	–	
Wallace	William R.			1901					34x36x12	not on original list
Walters	Marion E.	1		1911	A	14	4	Broadway	–	
Walters	Mabel	16		1904	G	8	6	Broadway	–	
Walters	Emma	62		1916	G	8	4	Broadway	–	
Waltrows	John S.	72			D		45	Broadway	–	
Ward	mary Agnes	1 mo		1915	7	11	3	Mount Olive	20x42x6 (as of the 1990 list)	
Wardman	Margarette May	39	1873	1912	7	23	4	Mount Olive	–	
Warfield	J.M.D.	Infant			5	13	3	Mount Olive	–	
Watts	America	82	1851	1934	5	21	6	Mount Olive	–	
Watts	Willie		1879	1909	5	21	2	Mount Olive	–	
Watts	Martha Jane	60	1874	1934	5	21	8	Mount Olive	–	

Sunday, August 02, 2009

Page 71 of 75 pages from the cemetery information concerning the burials showing William R. Wallace's name absent of location data.

A Statement From Dale Bybee, the Man Responsible for the Recovery and Donation of Colonel William Wallace's Headstone

My connection, or story, begins in December 2011. I was attending an outdoor event in Long Beach, California. This event is hosted by the George Gunther family. George Gunther was a lover of history and a preservationist. George has since passed, but his property continues to serve as a living museum. His collection showcases, single cylinder engines, crawler tractors, and many antique items. On his property there is a railroad spur where railroad speeders are operated. Anyone is welcome to show and run everything from cars, motorcycles, trucks – you name it.

While attending the show I noticed a large tombstone with the base. Wanting to build a Western theme cemetery on my own property, I asked if it was available. The family had no personal ties, other than their father George had acquired at some point. This was not the Wallace stone.

Some time had gone by and I was contacted by one of George's sons, Bruce. He said he began to wonder where this stone had come from. He wondered where and how his family had acquired the stone. His search led him to a person at a local museum and a possible connection. Bruce gave me the number of a person that worked at a museum that might have access to more tombstones. I made arrangements to meet where a number of stones were stored.

There were probably 130 stones piled 5 feet to 6 feet high. The wood separating them had long since rotted away. They needed the stones moved from where they had been stored for over 15 years.

I'm now faced with the prospect of acquiring some or all of the stones. I say stones, as there are 40% that appear to be only what would be the base portion with no name or dates. Through my research I found that during the re-internment, that the top and bottom did not remain together. Whatever the reason, here they were. Perhaps I need to start at the beginning. How did these tombstones come to rest in an outdoor storage area covered the with years of dust? After the two cemeteries had been made into a park, one being Mount Olive, the other Broadway cemetery, each portion of the park that represented the cemetery named had a granite and concrete monument erected. In front of each monument is a dedication plaque listing days of interment. It would appear that William Wallace was buried at Broadway Cemetery. The Broadway Cemetery is bordered on the west side by Citrus Avenue and Dorland Street on the north side. Broadway forms the southern boundary.

Mount Olive is on the west side of Citrus Avenue. They are all unique and had historical significance as far as I was concerned. The dash between the dates here on earth meant something. The decision was made, all would be removed and placed on land in Action, California. As I found out later, one stone was known to serve as a mailbox post. Another one, a base, had rope tied around it and was raised and lowered to compact trash. Somehow, through the years, the tombstones were scattered around, acquired, perhaps without even knowing what they were (in the case of the base portion) or where they had come from. After they were trucked, four fifty-foot truck loads later, some were individuals that fought in the Civil War. Some sadly had a last name with "Baby" proceeding it. No first name.

Dale Bybee in front of a headstone in his private cemetery, which consists of discarded headstones from the Mount Olive and Broadway cemeteries. *(Courtesy Dale Bybee)*

I was contacted in late 2016 by Chuck King. He asked if I had a certain tombstone. When he told me the name, I knew I had it without checking. The tombstone was of course William R. Wallace, founder of Wallace, Idaho. If this was the case, I knew this had to be placed in the town named after him. As an antique collector, to have a piece with providence is truly special. As it turned out, this was indeed the founder of the town of Wallace Idaho. Arrangements were made to pick up the tombstone and transport it to the town of Wallace.

Headstones without bodies. Dale Bybee's new cemetery located at the intersection of Sierra Highway and Red Rover Mine Road in California. *(Courtesy Dale Bybee)*

Jamie and Barbara Baker Drove to California to Transport Colonel William Ross Wallace's Discarded Headstone to Its Permanent Place of Honor in Wallace

Some of the people involved in making Colonel Wallace's headstone available as a permanent memorial at the Northern Pacific Railroad Depot Museum & Gift Shop in Wallace, from left: Dave Copelan, coordinator, Wallace Idaho Chamber; Shauna Hillman, museum director; Barbara and Jamie Baker, owners of the Red Light Garage and the Hercules Inn; and Chuck King, historian and one of the founders of the Spokane Valley Museum located at 12114 East Sprague Avenue, in the Spokane Valley. Though it's not as visible in this photo as the one on page 13, etched on Wallace's headstone is an insignia of the Knights Templar, an ancient branch of the Masonic order, which is credited with being the founders of the military orders. *(Bamonte photo)*

William Ross Wallace Obituaries

William Wallace's obituary from the November 17, 1901, *Los Angeles Times*

One of Those Who Developed The Great West

Was a Prime Mover in Coeur d'Alene District, Operated in Arizona and Mexico, and Lived in Los Angeles – Gen. Lew Wallace's Cousin.

Col. W. R. Wallace died at 8:30 last night at Hotel Greenleaf, in Whittier, where be had been staying since the 7th. The cause of death was Bright's disease from which he had been a sufferer for some time. He was 67 years of age.

Col. Wallace was one of the best-known mining men in the West. He was one of the movers in the opening of the Coeur d'Alene district in Northern Idaho, and founded the town of Wallace. He had also been interested in mining operations in Canada, Minnesota, Texas, New Mexico, Mexico and Arizona, and lived for a number of years at Prescott. On account of failing health he came to Los Angeles about a year ago from Arizona.

.

Col. Wallace was born at Lexington, Ky. February 26, 1834, and was cousin of Gen. Lew Wallace.

He was educated at Amherst College. He served with credit in the war for the Union, entering the service in the Thirteenth Ohio Infantry. Later he served on the staff of Gen. Rosecrans, and at the close of the war was in command of the Second Kentucky Cavalry. He was a Mason and a life member of Albany Commander No. 22, Knights Templar of Albany, N. Y.

He leaves a widow and three children, one of whom is the wife of Robert Griffith of Los Angeles. Mrs. Wallace and Mrs. Griffith were at the bedeside at the time of his death. The funeral will be held at Masonic Hall, Whittier, at 1 p.m. Monday, and the interment will be in the Whittier Cemetery, according to the wish of the deceased.

William Wallace's obituary from the November 22, 1901, *Whittier Register*

A Prominent Man Passes Away at the Greenleaf – Decease of Col. W. R. Wallace

The friends of Col. W. R. Wallace were pained early Sunday morning by the announcement of his death, which occurred on Saturday evening.

Col. Wallace came here with his wife, daughter and son-in-law only a short time ago and they were guests at Hotel Greenleaf. He had been very ill for some time of what was thought to be Bright's disease, but a post-mortem examination revealed as the cause of death, an attack of appendicitis, from which he had also suffered years ago.

Col. Wallace served honorably in the Civil War, and has since been a prominent mining man. He is the cousin of Gen. Lew Wallace, the author of Ben Hur. The funeral took place in the Mason's Hall on Monday at 1 P. M. The services were conducted by N. D. Ellis, W. M. of Whittier Lodge. The Hall was beautifully decorated by the ladies of the Eastern Star. There were a number of handsome floral pieces sent from afar by friends of the family.

The remains were interred with Masonic honors in the Whittier Cemetery, the pall-bearers being, Judge Madison T. Owens and Mr. O. M. Souden of this city, and Hon. W. A. Harris, George Kissenbury, H, H. Metcalf, J. R. Newberry, Dan McFarland and G. J. Griffith of Los Angeles. The deceased leaves a wife and three children to mourn the loss of a devoted husband and father.

Card of Thanks

We, the widow and daughter of William Ross Wallace, do sincerely thank the Whittier Lodge F. and A. M., the Ladies of the Eastern Star, the veterans of the Grand Army, and all other friends who so kindly assisted at the funeral, and during the last sickness of our beloved, and assure them that their kindness and sympathy are fully appreciated.

ANNIE E. WALLACE – ANNIE G. GRIFFITH

Los Angeles Times

November 17, 1901

pg. 8

FUNERAL OF COL. WALLACE.

Noted Western Mining Man Laid to Rest at Whittier—Mourned by Many Admiring Friends.

Under a pall of white carnations, white chrysanthemums and smilax, the remains of the late Col. W. R. Wallace were carried to the grave yesterday afternoon at Whittier. The funeral services, which were very impressive, were conducted under the auspices of the Whittier Masonic lodge. The pall-bearers were Hon. Will A. Harris, George Kissenbury, H. H. Metcalf, Dan McFarland, J. R. Newberry, all of Los Angeles, and Maj. Madison T. Owens of Whittier, warm personal friends of the deceased and his family.

Col. Wallace died at Hotel Greenleaf, Whittier, last Saturday evening, after an illness of considerable duration. It was thought that his malady

THE LATE COL. W. R. WALLACE.

was Bright's disease, but a post-mortem examination showed that death was due to the effects of an attack of appendicitis several years ago.

Friends of Col. Wallace of long standing say the dominant trait of his character was his unbounded generosity. A typical western mining man, he was the friend of every person with whom he had dealings, from humble mine laborer to multi-millionaire. His relations with his own family and close personal friends were most tender and affectionate. Among his warm personal admirers were Gen. Grant and other distinguished veteran officers of the Civil War. While mining in Colorado, Gen. Grant paid him an extended visit in his mountain cabin.

During periods of stress in the early history of the Coeur d'Alene mining district in Idaho, he kept many persons from starving by supplying them with the necessaries of life when they had no funds. The town of Wallace, Idaho, was founded by him, and its citizens always held him in grateful generosity. During his brief residence in Southern California he made many friends here, as was attested by the large attendance at his funeral.

Col. Wallace was a cousin of Gen. Low Wallace, and served with distinction on the staff of Gen. W. S. Rosecrans during the Civil War.

The following is an excerpt from the obituary in the *Los Angeles Times*, November 17, 1901

Friends of Col. Wallace of long-standing say the dominant trait of his character was his unbounded generosity.

A typical Western mining man, he was a friend of every person with whom he had dealings with, from humble mine laborer to multimillionaire. His relations with his own family and close personal friends were most tender and affectionate.

Among his warm personal admirers were General Grant and other distinguished veteran officers of the Civil War. While mining in Colorado, General Grant paid him an extended visit in his mountain cabin.

During periods of stress in the early history of the Coeur d'Alene mining district in Idaho, he kept many persons from starving while supplying them with the necessities of life when they had no funds. The town of Wallace, Idaho, was founded by him, and its citizens always held him in grateful generosity. During his brief residence in Southern California he made many friends here, as was attested by the large attendance at his funeral.

Col. Wallace was a cousin of Gen. Lew Wallace, and served with distinction on the staff of Gen. W. S. Rosecrans during the Civil War.

Idaho State Tribune

December 25, 1901 (excerpt)

excitement or stampede saw the old colonel among the early arrivals, he having mined in nearly all the western and southern states. To the Coeur d'Alenes he came during the fall of 1883, coming up the south fork of the Coeur d'Alene river. At that time the old Mullan road was completely blocked—burnt and fallen timber obstructing the road—and bridges had been destroyed by fire. He opened up the road, built bridges and trails up Nine Mile and Canyon creeks, and started the road which leads up Niae Mile from Wallace towards Murray aad later on, recovering the contract, completed the road across

COL. W. R. WALLACE.

the divide.

Realizing after the discovery of lead ore on his Ore-or-no-Go claim and the later important strikes on the Poorman nnd the Tiger lodes, that the point where the waters of of Canyon and Nine Mile creeks mingle with that of the south fork of the Coeur d'Alene river was destined to become a mining center the colonel concluded to locate here and fenced in 80 acres of land and gave to the early comers libearl inducements to locate, and thus established the town of Wallace.

In the spring of 1890, and shortly before the town of Wallace was

William Ross Wallace died on November 16, 1901
The Following Story Appeared in the *Idaho State Tribune* on Christmas Day, Wednesday, December 25, 1901.

WALLACE'S FOUNDER AND THE PRESENT MAYOR

Colonel W. R. Wallace, the Founder of the City and P. F. Smith, its Present Executive Head.

This Article is Significant Because it Contains One of Only Two Photographs I Was Able to Find of William Ross Wallace.

We have been successful in obtaining for this issue of the Tribune a splendid picture of the father of the city of Wallace, the late Colonel W. R. Wallace, [photo from this article is on page one] *who will be held in grateful remembrance by the old time pioneers and prospectors of the early days. They well know that there was no latch to his cabin door, his place being the headquarters for all prospectors, and those who know him best say there was no more kindhearted nor more liberal man than old Colonel Wallace—ever ready to share his grub and his blankets with those who came to his cabin door.*

Colonel Wallace was born in Lexington, Kentucky, February 14, 1834. He served through the Civil War and was wounded several times.

Ever since the early sixties he followed mining and nearly every excitement or stampede saw the old colonel among the early arrivals, he having mined in nearly all the western and southern states. To the Coeur d'Alenes he came during the fall of 1883, coming up the south fork of the Coeur d'Alene river. At that time the old Mullan road was completely blocked – burnt and fallen timber obstructing the road – and bridges had been destroyed by fire. He opened up the road, built bridges and trails up Nine Mile and Canyon creeks, and started the road which leads up Nine Mile from Wallace towards Murray and later on, recovering the contract, completed the road across the divide.

Realizing after the discovery of lead ore on his Ore-or-no-Go claim and the later important strikes on the Poorman and the Tiger lodes, that the point where the waters of Canyon and Nine Mile creeks mingle with that of the south fork of the Coeur d'Alene river was destined to become a mining center the colonel concluded to locate here and fenced in 80 acres of land and gave to the early comers liberal inducements to locate, and thus established the town of Wallace.

In the spring of 1890, and shortly before the town of Wallace was consumed by fire, Colonel Wallace broke up camp and again followed mining, first in the Lake Superior country, then in Arizona and a large section of New Mexico, and afterwards settled down in California, where he found his resting place. We clipped the following obituary from a San Francisco paper of November 18 [date conflicts].

"Colonel Wallace died at Hotel Greenleaf, Whittier, last Saturday evening, after an illness of considerable duration. It was thought that his malady was Bright's disease, but a post-mortem examination showed that death was due to the effects of an attack of appendicitis several years ago.

"Friends of Colonel Wallace of long standing say the dominant trait of his character was his unbounded generosity. A typical western mining man, he was the friend of every person with whom he had dealings, from humble mine laborer to the multi-millionaire. His relations with his own family and close personal friends were most tender and affectionate. Among his warm personal admirers were General Grant and other distinguished veteran officers of the Civil War. While mining in Colorado General Grant paid him an extended visit in his cabin.

"During periods of stress in the early history of the Coeur d'Alene mining district in Idaho, he kept many persons from starving by supplying them with the necessaries of life when they had no funds. The town of Wallace, Idaho, was founded by him, and its citizens always held him in grateful generosity. During his brief residence in southern California he made many friends here, as was attested by the large attendance at his funeral.

"Colonel Wallace was a cousin of General Lew Wallace, and served with distinction on the staff of General Rosecrans during the Civil War."

His son, Oscar B., is well known having been with his father the greater portion of his time in this district, mining and operating with him, and later on independently. During the years of '92 and'93, he was mayor of this city. At the present time he is operating in the vicinity of his old stomping ground, west of the Black Cloud and California mines, for the Ruth Mining company, which is principally owned by an eastern syndicate of capitalists.

William Ross Wallace's last stop
Whittier, California
A Short History of the City of Whittier
(From the city's website)

Whittier, California, began when Jonathan Bailey and his wife, Rebecca, became two of the first residents. The Baileys closely followed the Quaker religion, holding regular religious meetings on their front porch. This devotion to their religion began setting the tone for the Quaker religion with other newcomers to the area. As the city grew, the citizens named it after John Greenleaf Whittier, a respected Quaker poet.

Whittier is located in Los Angeles County, California, about 12 miles southeast of Los Angeles. As of the 2000 census, the city had a total population of 83,680. It is the home of Whittier College. Like nearby Montebello, the city is considered to form part of the Gateway Cities and the southeast area of the county.

The roots of Whittier can be traced to a Spanish soldier, Manuel Nieto, who in 1784 received a Spanish land grant of 300,000 acres as a reward for his military service and to encourage settlement in California. Nieto's area was reduced in 1790 because of a dispute with the Mission San Gabriel, but he still was left with 167,000 acres stretching from the hills north of Whittier, Fullerton and breaking, south to the Pacific Ocean, and from today's Los Angeles River, east to the Santa Ana River. Nieto constructed a dwelling for his family near the present town of Whittier, stocked the land with cattle and horses, and cultivated corn. Upon his death in 1804, his children inherited his property. At the time of the Mexican-American War, much of Whittier was owned by Pio Pico, a rancher and the last Mexican Governor of California. Pio Pico built a hacienda in Whittier on the San Gabriel River, which today is Pio Pico State Historic Park

After the war, Jacob F. Gerkens, a German immigrant who paid $234 to the U.S. government to acquire 160 acres of land (under the Home-

stead Act) built a cabin on the land which is today known as the Jonathan Bailey House. Gerkens would later become the first Chief of Police of the Los Angeles Police Department. The land had several owners before a group of Quakers bought the land, which had since been expanded to 1,259 acres, with the purpose of founding a Quaker community.

The area soon became known as a thriving citrus ranching region, with "Quaker Brand" fruit being shipped all over the United States. Soon after, walnut trees were planted, and Whittier became the largest walnut grower in California.

In 1888 The Southern Pacific Railroad completed its first line to Whittier. At that time Whittier pioneers built the Greenleaf Hotel.

Colonel Wallace's Famous Dog at Whittier, California

The following article appeared in the *Los Angeles Times* on March 12, 1907, on page 17. It was titled, "Farewell, Faithful Companion. San Juan Dog Hero Falls Before Auto":

After having braved the leaden ball of that deadly rush up San Juan Hill with the Rough riders fearless in the face of the enemy and after having passed through dangers that made cowards of brave men, Don, a Cuban bloodhound, tho best-known and probably the best-loved dog in the Whittier community, died in that city yesterday.

He was the victim of a scorching automobilist. Don was the property of Dr. F. H. Hadley, and was valued both because of his history and well-tested, bravery, and for an almost human intelligence.

He was given by Col. Roosevelt, in the stormy days of the Cuban campaign, to his lieutenant, Hamilton Fish, Company B, of the Rough Riders, and became the pet of the soldiers and one of their mascots. Pressing close to his master, Don took his part in the famous rush up San Juan Hill, and woe to the unlucky mortal who opposed him. His great strength and unfaltering courage stood him in good stead that day, and when the boys who took the hill tell of those awful hours, Don, their mascot, always comes in for a word of praise. ***But Hamilton Fish, his master, fell that day, and the faithful pet was given by his comrades to Col. William Wallace, whose property he was for several years.***

At the death of this owner [Colonel William Wallace] ***in Whittier, a few years ago, Mrs. Wallace presented Don to Dr. Hadley, the family physician.*** *Since that time the big dog had the freedom of the Quaker town, and had never walked through the streets without receiving much attention from small boys and girls to those of larger growth. The accident which caused his death occurred just at dusk. A big touring car containing four persons, going around a corner at so high a speed that the old dog, which was walking quietly along could not get out of its way. Everything possible was done by the members of the auto party, who seemed to feel keenly the effect of their careless driving.*

Theodore Roosevelt in 1898. *(Public domain)*

SAN JUAN DOG HERO FALLS BEFORE AUTO.

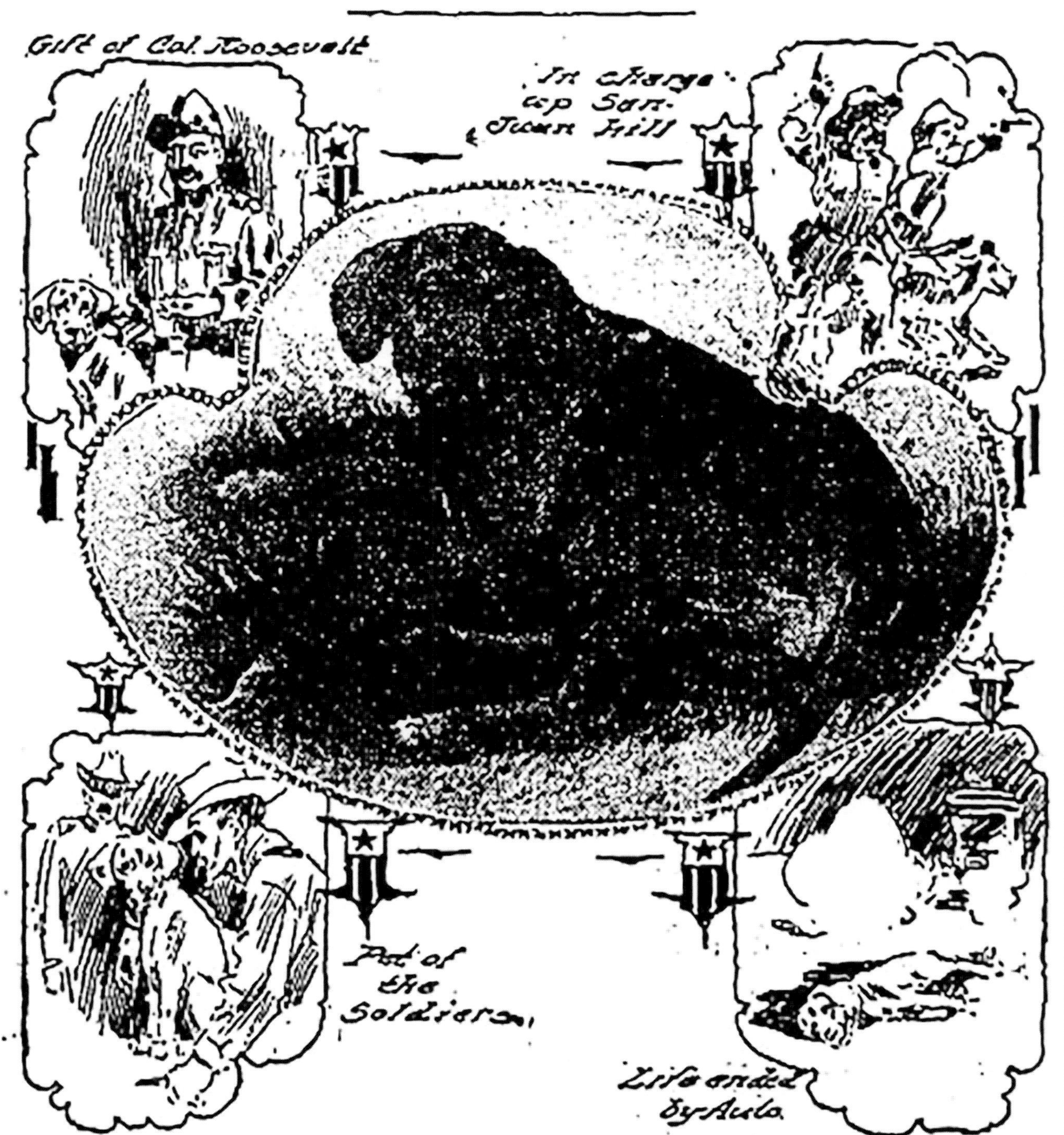

Only a dog. Noted canine which died at Whittier yesterday, and some of the incidents in its career.

Man's best friend.

From the March 12, 1907, *Los Angeles Times.*

Colonel William R. Wallace's Thoughts on Teddy Roosevelt Before Roosevelt Became President

COULDN'T BE ELECTED CONSTABLE IN ARIZONA

W. R. Wallace, a wealthy mine owner of Prescott, Ariz., is not an admirer of Governor Roosevelt. Mr. Wallace was a soldier for four and a half years in the civil war, was one of the founders of Leadville, and has had extensive experience in the West as mine manager and mine owner. He does not hesitate to denounce Roosevelt for pretensions which he declares are flaunted by the chief jingo orator of the Republican party.

"Roosevelt," said Mr. Wallace, at the St. James hotel last night, "has publicly claimed that all Democrats are cowards. He could not be elected to the office of constable in the territory of Arizona. Take his own book and you will see that he is not a fighter. He made no charge up San Juan hill. He went no farther than the block house and the negroes carried the hill. It was neither Roosevelt nor the alleged rough riders who swept the enemy off San Juan hill. I knew Roosevelt when he was a cowboy in Wyoming and I have watched his career ever since. He is always for Roosevelt. He has written five books on Roosevelt and what Roosevelt has done, and the books are all alike. Read one of them and you have read them all. Down in Arizona we despise him and to my certain knowledge fifty-three members of a Grand Army post in the town of Fort Collins, this state, have signed a pledge that they will vote against Roosevelt. These men are sick of the pretensions of Roosevelt and they know what made Colorado. They recognize that it was silver that made this state and this city what they are. Before the discovery of silver at Leadville I was offered the entire block opposite the Tabor opera house for $2,300.

"It is worth millions of dollars now. Thirty years ago the country in Northern Colorado through which Roosevelt traveled to-day was a barren plain. To-day it is the garden spot of the continent. It was the silver mines that made that country worth development. The farmers planted little patches of gardens and found a ready sale for their produce at the silver mines. They continued to extend their fields and the miners took all they raised. We should never forget the debt we owe to silver. I honor the men who stand up for the white metal."

Mr. Wallace is as warm a friend of Senator W. A. Clark as he is an enemy of Roosevelt. He has visited the United Verde mine in company with Senator Clark and speaks from personal knowledge of the great mine. He says the net income of the owner of the mine from the property is $1,000,000 a month.

This article appeared on page 5 of the Denver *Rocky Mountain News* on Thursday, September, 27, 1900. *(Public domain)*

The Three Wives of Colonel Wallace, and Other Family Members

Wife number one: Sarah or Sarrah Slade, born in New York in 1837. At the time of the 1850 census, she was 13 years old and living with her father, Servil Slade, and her mother, Fanny Slade, at Litchfield, Herkimer County, New York.

William R. Wallace and Sarah married in 1857, in N.Y. At the time of their marriage, William would have been 23, and Sarah 20. In 1860, were living in Alliance, Stark County, Ohio, and were the parents of their two-year-old, Oscar. William's occupation was as a merchant.

They had a daughter, Emma E., born in Ohio about 1861. By 1870, the family had moved to Titusville, Crawford County, Pennsylvania, where William was an agent for a sewing machine company.

The Wallace family appeared in the 1880 census for Centerville, Appanoose County, Iowa. Listed were William, Sarah, Oscar, and Emma, whose last name was now Lohnridian, with a son, Robert. By this time, Emma was divorced. William's occupation was mining.

In the Iowa State Census for 1885, Sarah, age 48, was living in Centerville, Appanoose County, Iowa, along with Emma. Emma had remarried, to a Mr. Neuse, but he was not living with her. Emma had two sons, Robert Neuse and Roy Neuse. The state census indicates that William Ross Wallace is no longer married to Sarah. She would have been married to William for the duration of his time in the Civil War, and up until sometime after 1880. Although no record of a divorce or last place of residence could be located, it appears William may have remarried sometime around 1883. William and Sarah would have been married for approximately 25 years.

A listing for Sarah Wallace was found in both the 1902 and 1903 *R. L. Polk & Co.'s Spokane City Directory.* She was living at E. 723 Mission Ave., the same address as her son, Oscar Wallace, and grandson, Walter Scott Wallace. In 1905, there was a listing for Oscar and Walter at 823 Sinto.

Apparently, based on the next record of Sarah, she moved to Mapleton Township in Blue Earth County, Minnesota. The record was her obituary, which appeared in the *Blue Earth County Enterprise* in 1904:

A Good Woman Called Home

The angel of death called at the home of John Goff Tuesday afternoon and relieved from her suffering Mrs. Sarah Wallace. During the winter of 1881 she and her daughter Emma lived in Mapleton. Early last fall Mrs. Wallace came from Denver, Colorado, to visit with her cousin, John Goff and family, shortly after arriving here she was taken ill and much of the time since has been confined to her bed. Death came as a relief.

Sarah Slade was born in Litchfield, Herkimer county, New York, March 1st, 1887, and at the time of her death was 67 years, two months and nine days old.

In 1857, she was married to William R. Wallace, and since that time has traveled a great deal. Two children were born to this couple, Emma Goss, of Denver, Colorado, who was with her mother at the end, and Oscar Wallace, of Spokane, Washington. A brother and two sisters are also still living in New York.

About 35 years ago Mrs. Wallace united with the Baptist church at Titusville, Pa., and her Christian training and gentle disposition made her a patient sufferer.

Funeral services were held at John Goff's residence, the remains being laid to rest in the Chase cemetery – Rev Bolton, of Mapelton, officiated.

Wife number two: <u>Lucille "Lucie" (alternately spelled "Lucy"), maiden name unknown,</u> was born in 1848. According to their divorce records, she married William Ross Wallace on July 22, 1871, in Syracuse, New York. However, there appears to be a typo, as he was still married to Sarah in 1880. Perhaps they married in 1881. William was married to Lucie during the time he founded the town of Placer Center (Wallace). The following passage from Richard Magnuson's book, *Coeur d'Alene Diary*, describes her trip to Wallace in 1885:

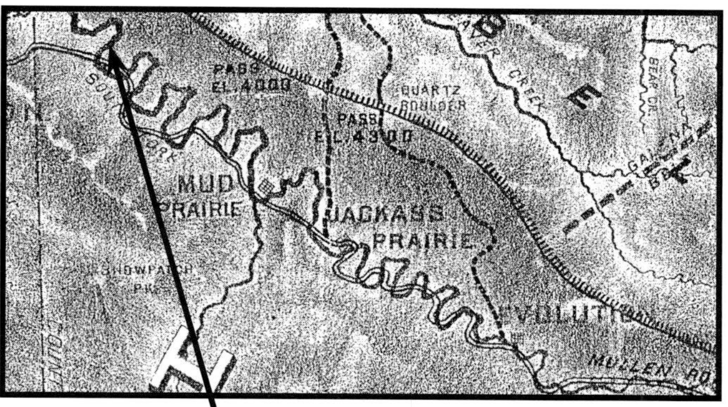

The start of the zigzagging course of the South Fork of the Coeur d'Alene River, which his second wife, Lucie, describes. This map was compiled on January 25, 1884, by F. L. Miller, Civil Engineer, and titled: *Map of the Coeur d'Alene Mines and Vicinity, Idaho Territory.* It shows the fording of the South Fork of the Coeur d'Alene River as described by Mrs. Wallace on her first trip to Placer Center, where she states "After leaving the Mission they forded the river 14 times before arriving at Placer Center . . ."

Colonel Wallace's Wife Lucy Arrives at Placer Center

On June 22, 1885, Colonel Wallace drove a wagon to pick her up at the Old Mission. She and her luggage, which included a dog, bird, cats, and chickens, had arrived early in the day by boat. After leaving the Mission they forded the river 14 times before arriving at Placer Center, and this was the cause of some concern. The water was running high, but their only loss was one chicken which drowned when the water dashed over the wagon box. The 25 mile trip took two days, with an overnight stop at Jackass Prairie.

At the time of Mrs. Wallace's arrival, the town's population numbered 14. She was the only woman to "winter" in Placer Center that year. There were two women in Burke, Mrs. F. R. Culbertson, a daughter of S. S. Glidden, and a French woman. The former could not speak French and the latter could not speak English, so there was not much conversation between them. A woman also "wintered" that year at Jackass Prairie. In August, 1886, Lucy Wallace was appointed the first postmistress in town. She later stated that the United States Post Office Department would not accept the name of Placer Center because it was too long. Her husband objected to a town's being named after him because there were many towns of this name already. Lucy Wallace ignored her husband's opinion and filled out the United States Postal Department slip listing the Post Office in the name of Wallace, and so it has remained to this day.

Obituary for Mrs. Lucille A. Wallace, first woman resident of Wallace

The following article appeared in the May 23, 1929, edition of the *Wallace Miner*:

In Seattle last Wednesday Mrs. Lucy A. Wallace, first woman to arrive in Wallace, died at the age of 81. Mrs. Wallace was the wife of the late Colonel Wallace, founder of this city and in whose honor it was named. Colonel Wallace built a commodious log cabin at the head of what is now Sixth street, occupying a bench that was reduced for the construction of the building now occupied by the C. Z. Seelig company, where he and Mrs. Wallace resided in the pioneer days. Mrs. Wallace owned in her own right the corner now occupied by the Smoke House, where she erected a two-story frame building. In the corner room she conducted a store and when the post office was established in 1886 she was appointed postmaster, the office being in her store. The Seattle account of her death states that she left instructions to chloroform her dogs upon her death, fearing that they would not receive kind treatment when she was gone. It is recalled that she was very fond of dogs when she resided in Wallace, one called "Tip" being her constant companion. ***Mrs. Wallace is kindly remembered by residents of Wallace of that day, although the conflict between the citizens and Colonel Wallace over the title to the townsite caused an estrangement which was the real cause of her removal from Wallace.*** *About a year ago she sent to the board of trade her commission as the first postmaster of Wallace to be preserved as a souvenir of the early days.*

Another interesting obituary appeared on Monday May 20, 1929, in the *Cleveland* (Ohio) *Plain Dealer*:

MA PETTINGILL OF GOLD CAMPS DIES

[Lucie Wallace was a famous character in the book *Ruggles of Redgap*, written about her life at Cripple Creek]

Negro Servant to Bury Woman Prospector and Then Kill Dogs.

Seattle. May 19. — Bowed with sorrow, Bill McBrown negro, today prepared to bury Mrs. Lucy Wallace.

She had known wealth, but she died in poverty. Cripple Creek, Coeur d'Alene, the Yukon trail, places of romance, where men rushed, and toiled for gold—knew Mrs. Wallace.

The nation knew her in "Ma" Pettingill, for she was said to have been the original of Barry Leon Wilson's famous character in "Ruggles of Red Gap." Wallace, Ida., was named for her husband.

She died Wednesday alone except for McBrown and her attorney, Phil Tworoger.

After her funeral McBrown will kill her two faithful old dogs. "Shoot the dogs. Bill," she said, as death neared. "Shoot them right away when I die. No telling what kind of care they'd get after I'm gone. [author's emphasis]

Born in New York

Tworoger recalled part of her adventurous life today. Mrs. Wallace was born in Mohawk, Valley, N. Y., in 1848. She was reputed to have been the first white woman at the Cripple Creek gold camp and at Coeur d'Alene, Id., where she went with her husband, a Civil War veteran.

Col. Wallace died in Coeur d'Alene, about 1890 [This is incorrect. He died in 1901, though she was listed in the 1896 Spokane directory as a widow.] *Tworoger said, and Mrs. Wallace was forced to earn her own living. She went to Spokane, where she was said to have been the city's first telephone operator. In 1898 she tramped the Yukon trail, but in 1901 she returned to Seattle and hired Mr. McBrown as her servant, and for 28 years he worked for her.*

Poverty approached with old age. Friends drifted away, but McBrown remained faithful. He cared for her and provided for her,

To Bill, Mrs. Wallace left her cottage, and to Mrs. Marie Burke, secretary to Tworoger, she bequeathed her piano.

Lucille Wallace was buried at Seattle's Evergreen-Washelli Memorial Park.

Colonel William Wallace's wife, Lucie, was the town's first postmaster

In his book, *Coeur d'Alene Diary*, Richard Magnuson described the following:

A Town Called Wallace, 1884

A newspaper printed at Eagle City carried the birth announcement, in its May 10, 1884, issue:

"PLACER CENTER — This is the name of a new town started on the south fork of the Coeur d'Alene, at a point about seven miles up the road from Evolution. This town is situated in a good location and commands Canyon Creek mines and other tributaries of the South Fork, wherein mining in a small way is going on. The town will be a good point for prospectors who intend to put in a summer's work on the range between the Coeur d'Alene and St. Joe, and its permanency is assured from the fact that it is on

Ruggles of Red Gap

By Harry Leon Wilson, as advertised in Amazon

A humorous adventure about a British valet out of his element in a small Western town in the USA. Ruggles becomes the servant of an American couple after his employer uses him as a stake in a poker game. He travels to the West and is mistakenly identified as a wealthy Brit. He becomes a small town celebrity with many humorous twists in the plot.

Ruggles of Red Gap was initially serialized beginning December 26, 1914 in The Saturday Evening Post and became a best selling novel in 1915, adapted for the Broadway stage as a musical the same year and made into a film several times, most famously in 1935 starring Charles Laughton. Harry Leon Wilson: Some Account of the Triumphs and Tribulations of an American Popular Writer.

Some Account of the Triumphs and Tribulations of an American Popular Writer. By George Kummer. (Cleveland, Ohio: The Press of Western Reserve University, 1963

The author suggests in the Preface two reasons for having written a biography of Harry Leon Wilson: first, it will serve as a reference for students of American literature, particularly American humor; and second, he hopes it will rescue this "popular writer" of the early 1900's from falling into obscurity. Dr. Kummer believes that the creator of such delightful characters as Bunker Bean, Ma Pettengill, Ruggles, Cousin Egbert, Professor Copplestone, and Merton Gill deserves to be remembered and enjoyed by readers of the modern generation.

Wilson began his career as a writer at an early age in his father's newspaper and printing office in Oregon, Illinois. Later he was employed by the Bancroft History Company to assist in the collection of materials for the company's multivolume historical series. During his free time Wilson taught himself the art of writing, using as a guide the country's foremost humor magazine, Puck. In 1892 Wilson joined the staff of that magazine and four years later assumed the editorship.

Wilson resigned his position with Puck in 1902 and retired to his wife's family home in Missouri to write. After the completion of three mediocre novels, he collaborated with Booth Tarkington in writing plays for provincial America. The first and most successful of these was The Man from Home. As interest in the "road" theater began to decline and competition from Broadway became greater, Wilson settled in Carmel and returned to writing comic novels. Here, between 1912 and 1925, he produced his best works: Bunker Bean, Ruggles of Red Gap, Merton of the Movies, Oh, Doctor! and Professor How Could You!

Kummer's brief biography of Wilson is an excellent example of what can and should be done with minor American writers. Biographical details are kept to a minimum, and are generally concerned with events which most influenced his writings. As it should be, the major portion of the book is devoted to an analysis of Wilson's novels, the main aspects of which are very well handled. Kummer has made extensive use of Wilson's personal papers and has included a bibliography of his books.

The book should have a particular appeal for readers interested in the social history of the United States for, although his ability to write was limited, Wilson's novels provide a commentary on life in the 1920's. These views, frequently differing from those presented by other writers of the period such as Sinclair Lewis or F. Scott Fitzgerald, have been more than adequately explored by the author. Wilson, the defender of the small town, was a product of provincial America who wrote for provincial America. For this reason alone, Kummer is more than justified in his attempt to preserve Wilson's name for posterity.

the Mullan Road, which is the main emigrant road on the Bitter Root divide.

In retrospect, this is a very concise statement as to why, when, and where the city now known as Wallace, Idaho, came into being. The article fails, however, to note the journalistic requirement of reporting on the who's involved. It is the purpose of this writing to tell about the people involved in founding a city and a mining district, which were to become a center for mining for at least the next 100 years.

What starts a town? In the case of Placer Center, it was the building of a cabin by Colonel W. R. Wallace, to serve as his headquarters for mining exploration in the immediate area. He was a leader of men, and his group used the new townsite as their base camp. His cabin was built on the toe of a snow-covered hillside (near the south end of Sixth Street in Wallace). It was not until a year later that the Colonel's wife, Lucy, made her first appearance in Placer Center.

William and Lucie Wallace divorce in 1890

On January 22, 1889, after seven to nine years of marriage (the date of their marriage is unknown), Lucie Wallace began an action for divorce from William. She alleged he willfully abandoned her without cause and lived apart from her against her wishes.

Upon examining the divorce paperwork Lucie Wallace had generated for her divorce from Colonel Wallace, her husband had said nothing negative about her, and there appears to have been no property settlement. It is also noteworthy that both William and Lucie Wallace used the same attorney, William T. Stoll, for their divorce. Stoll done work for the Wallaces a number of other times, during a period of two years, when an attorney was required for other business matters they were involved in. This is significant for the fact that William was asking for no property settlement, or anything for that matter. He had made a decision that he did not want to live with Lucie any more, but did not give any further reason. Most notable, it did not appear that he either scorned or disrespected her in any way.

In the divorce decree, Lucie mentioned "I was very much afraid of him," yet she does not mention any reason for this fear, which is not typical in any type of divorce action. However, Annie Wallace, who William married two years after his divorce from Lucie, made two separate statements following his death, by which time they had been married for nine years. She said that Colonel Wallace was ***"a fine mining man and had many fine traits,"*** and ***"My relatives were not crazy about my getting married to anyone again, though Col. Wallace was a perfect prince."***

It is concerning that she gave orders to kill her dogs upon her death. In my opinion, that doesn't reflect well on her character. This not only seems heartless but also irresponsible. It's upsetting to even think about. Why would anyone want to end the lives of their precious pets upon their death instead of finding homes for them? There are many people in the world who would have gladly given them a good home if they knew the circumstances. It is also troubling to imagine the feelings of the person to whom she gave the order to kill her two faithful old dogs. ***"'Shoot the dogs. Bill,' she said, as death neared. 'Shoot them right away when I die.' "*** She was asked demanding this of Mr. Bill McBrown, her servant who had been loyal to her for 28 years.

Returning to the issue of their divorce proceedings, she further alleged that on:

> *June 27th, 1890, the defendant* [William Wallace] *was duly served with a summons according to law, that he failed to appear, answer or demur thereto, his legal time for doing so having expired and the default of the defendant having been duly entered.*
>
> *Conclusions of Law: The plaintiff is entitled to a decree of this court dissolving the bonds of matrimony now and heretofore existing between the plaintiff and defendant, together with judgement for cost of this suit.*
>
> *Wherefore by virtue of the law and by reason of the premises:*

Some of the people behind the scenes.
Without the people in charge of the records within a county jurisdiction and their willingness and friendly help, it would be impossible to write any accurate record of the actual events and the dates important to their county's history.
This page includes three of those people at work in their offices, with our appreciation for the help they so willingly provided.

John Amonson, then director of the Wallace Mining Museum, helping with research for this book and our *Coeur d'Alene Gold Rush and Its Lasting Legacy*. *(Bamonte photo)*

Marla Anson (left) and Debbie Ruggles, two Shoshone County Courthouse employees who graciously helped with my research for this project. *(Courtesy Jamie Baker)*

It is ordered, Adjudged and Decreed that the marriage now and heretofore existing between the plaintiff, Lucie A. Wallace, and the said defendant, William R., Wallace, be and the same is hereby forever dissolved, and the said parties and each of them is hereby freed and absolutely released from the bonds of matrimony and the obligations thereof. That the said plaintiff do have and recover of and the defendant her costs herein expended duly taxed at $. . .

Done in open court at Osborn, Idaho, this 26th day of July.

Willis Sweet, Judge

The following was attested to by William T. Stoll, Lucie's attorney in a deposition of William Wallace:

Q. Do you know the plaintiff and defendant to this action?

A: Yes. Have known both intimately for two years.

Q: What do you know about defendant's desertion of plaintiff?

A: I have had frequent conversations with him concerning the same and in each instance he has told me that he would never again live with the defendant and that he had left her for good.

He told me also that he left the defendant on 18, January, 1889, after due deliberation, and that at that time he had thoroughly made up his mind to never again live with her and that he had never since changed his mind.

He said also that when he left Idaho on January 18, 1889, or thereabouts that he would never have returned to Idaho had not business matters called him back.

Lucie eventually moved to Spokane, where she was listed in the 1896 Spokane City Directory as: "Wallace, Lucie A (wid Wm R) chief long distance opr Inland T & T Co, rms Cushng blk." (It was common in those days for the directories to indicate that a divorced woman was a widow.) As recounted earlier, she died in Seattle in 1929.

Wife number three: Annie E. Jones was born in 1848 at Elmira, New York. She married William Ross Wallace in 1892. The marriage ceremony was performed by Paster W. O. Lattimore, pastor of the Presbyterian Church. There were married until his death, nine years later, in 1901. Annie was 27 years younger than William. According to their marriage records, like her husband, Annie had been married twice before.

At the time of their marriage, William was living in Tacoma, Washington, and was involved in a mining operation. According to the "Record of Returns of Marriages in Marshal County, Indiana," William's father was named Lewis and his mother May. Annie's parents were Elijah and Emma.

Although it is unknown how long William and Annie knew each other before getting married in 1892, they probably met in Spokane. Annie may have made a trip to Spokane during the time William was working there. As mentioned earlier, William Wallace relocated to Spokane after his Wallace townsite property was jumped and Lucie had filed for divorce. There he engaged in business with General A. P. Curry. The 1890 *Spokane Falls Directory* listed them under the category "Real Estate, Mines, Stocks and Securities" as "Wallace, Curry & Co. (Wm. R. Wallace, A. P. Curry, Oliver Durant, ME Lindsey) Mining Brokers, 15 to 18 Tull blk." The following year, the 1891 directory included the listing: "Wallace Wm. R. (Wallace, Curry & Co.) mining brokers, 15 to 18 Tull blk." No other Spokane listings were noted.

For the next several years, William continued working in the mining industry, first in the Lake Superior country and later in Arizona and New Mexico. His final move to California was for health reason. As recounted earlier, he died there at age 67.

Filing for widow's pension

On September 4, 1928, 63 years following the end of the Civil War, a claim for a widow's pension was filed by Annie E. Wallace. At the time of her filing, Annie was living at #18 Edmund Avenue, Toronto, Ontario, Canada.

Another portion of a letter of interest from Annie Wallace

fine mining man &
had many fine traits
Oh! after I was Colonel
Wallaces wife for ten
years it seems odd
that I am asked to prove
that - which is easy to
do - but I was his
3rd wife - and they
wish me to tell where
the first wife died &
was buried & I dont

The ladys name was -
then they want the
record of his record

The above, is a portion of a letter to Royal S. Copeland, the secretary for a United States senator in Washington, D.C. Annie's handwriting was extremely difficult to read and many words were undecipherable. Of interest, on March 19, 1929, Annie wrote ". . . fine mining man and had many fine traits Oh! after I was Colonel Wallace's wife for ten years it seems odd that I am asked to have that, which is easy to do, but I was his 3rd wife, and they wish me to tell where the first wife died & was buried & I don't even know what . . . the lady's name was ?? then they want the record of his second wife's marriage and her divorce for W. R. Wallace before I married him! and his signature I've lots of papers & letter signed by him. If you will be so very kind as to see Mr. Winfrild Scott and tell him that I hope he . . ."

Annie's application was found in the National Archives under Dependence Original Number 1623115, case number 3711, bundle 63. It was addressed to the "Widow Division W. Oh. 1623115 regarding William Ross Wallace of the 13th Ohio Infantry." Annie's application contained 85 pages of correspondence generated in her attempt to obtain a Civil War widow's pension.

Rejection of widow's pension

Annie Wallace's widow's pension application was rejected. It was designated WO#1623115 and was based only on service in Co. H, 13th Ohio Infantry (three-month term). One of the reasons for being rejected is that Annie was unable to provide the actual dates of Col. William Ross Wallace's service during the Civil War. The information was apparently unavailable at that time, but some records have since become available (as discussed in Chapter II), but they are only from his first tour of duty, the three-month term referenced above.

As I learned during my research, official records from the most significant period of his service (the remainder of the war) are largely missing. However, it is known that he served under Major General Rosecrans, and apparently for a time on his staff, which Annie apparently did not know.

Some of Annie Wallace's correspondence in her endeavor to get the widow's pension

As can be seen from a sample of Annie's writings, they were exceptionally hard to read, were not well-worded, and full of spelling errors. Though challenging to decipher, the information contained in her correspondence is highly important to many of the events concerning William and her. In total, there are 79 pages of material from Annie.

In a letter written on September 4, 1928, to Mr. Winfield Scott of the "Widows Division," Annie stated the following:

My dear Mr. Scott, your kind letter so December 29 duly received. You certainly have been kind and helpful in writing – yourself. I am sure of being able to furnish showing statements as to the most important questions, but really I do not even know the name of W. R. Wallace's first wife, as he seldom discussed his two former wives".

I was of course with my husband when he died. I have many of his letters signed for the last one thousand dollar check he made out to me, which I never cashed."

I could have this check photographed and send you, or shall I send one of his signatures? Course I couldn't give his signature when he joined the Civil War as I never laid eyes on him [until] I went on a visit to Spokane falls Washington where I met him first. I never had any children by W. R. Wallace. I have one daughter.

Sometime after the death of W. R. Wallace in 1901, the representative for the "Widows Pension Division," who had been working her case, found out she had remarried following William Wallace's death and had withheld that information. When they asked her about it, the following was Annie's reply, taken from a letter to Mr. Scott on June 12, 1929:

I meant to have told you that, while on a visit to a new Goldfield, I married mining engineer Jack W. Minnis. I got married to him after I only knew him a week. As our marriage was in name only I sort of hated to mention it. He went away to seek his fortune in Alaska and on the 20th of September was killed by 10 cases of dynamite at his gold mine exploding. Only little bits of him were found.

The Hon. John M. Holzworth of White Plains New York, who was district attorney of Winchester County, and who visited Jack Minnis's camp the day after he was killed and who has to return in June to Alaska as a witness against the man he thinks did the killing of Jack.

I'm sure Mr. Hollingsworth will be only too glad to give a sworn statement of Jack's tragic end and my marriage to him had been annulled in my name restored as if I had never been married to Jack at all. Mr. Scott thank you for your kind advice to me and do hope how much you will be able to grant the pension, which I would like to have as I can't live many years more.

With all good wishes I am sincerely yours, Annie Wallace, 18 Edmond Ave., Toronto Ontario.

Another letter dated January 16 , 1929, to Senator Copeland from Annie Wallace stated:

My relatives were not crazy about me my getting married to anyone again, though Col. Wallace was a perfect prince. He was wounded and had a piece of shell in his head which the doctors at Whittier advised me to remove when they cut him to see what caused his death I wanted to know what ever he died of which now I had let them take out the pieces shell which it was thought. Dr. Robert B. Griffins, 3305 Wilshire Blvd. was present when W. R. Wallace had the autopsy performed to find out what he died of. I was sorry I hadn't had the doctors cut out the piece of shell his head he got in battle during the Civil War.

In a June 14, 1932, partial letter Annie Wallace wrote to Mr. Fletcher from the Widows Division, apparently quoting an article about her husband written in a newspaper following the Civil War:

"... was riding off of the battlefield and I saw Colonel lying on the field, left for dead and had him removed and he like ???? & ???? at a hotel in Cincinnati or he was wounded in Ohio..."

All of Annie Wallace's correspondence to the widow's Pension Division stopped in 1932.

In regards to the last days or demise of Annie Wallace, my researcher stated she had used every resource and genealogy tool available, but was unable to find any more information about her. She found pieces of information on Annie's daughter (Colonel Wallace's stepdaughter), Ann G. (Bullene) Norrington, and her family, but nothing that shed any light on Annie's whereabouts or her subsequent death.

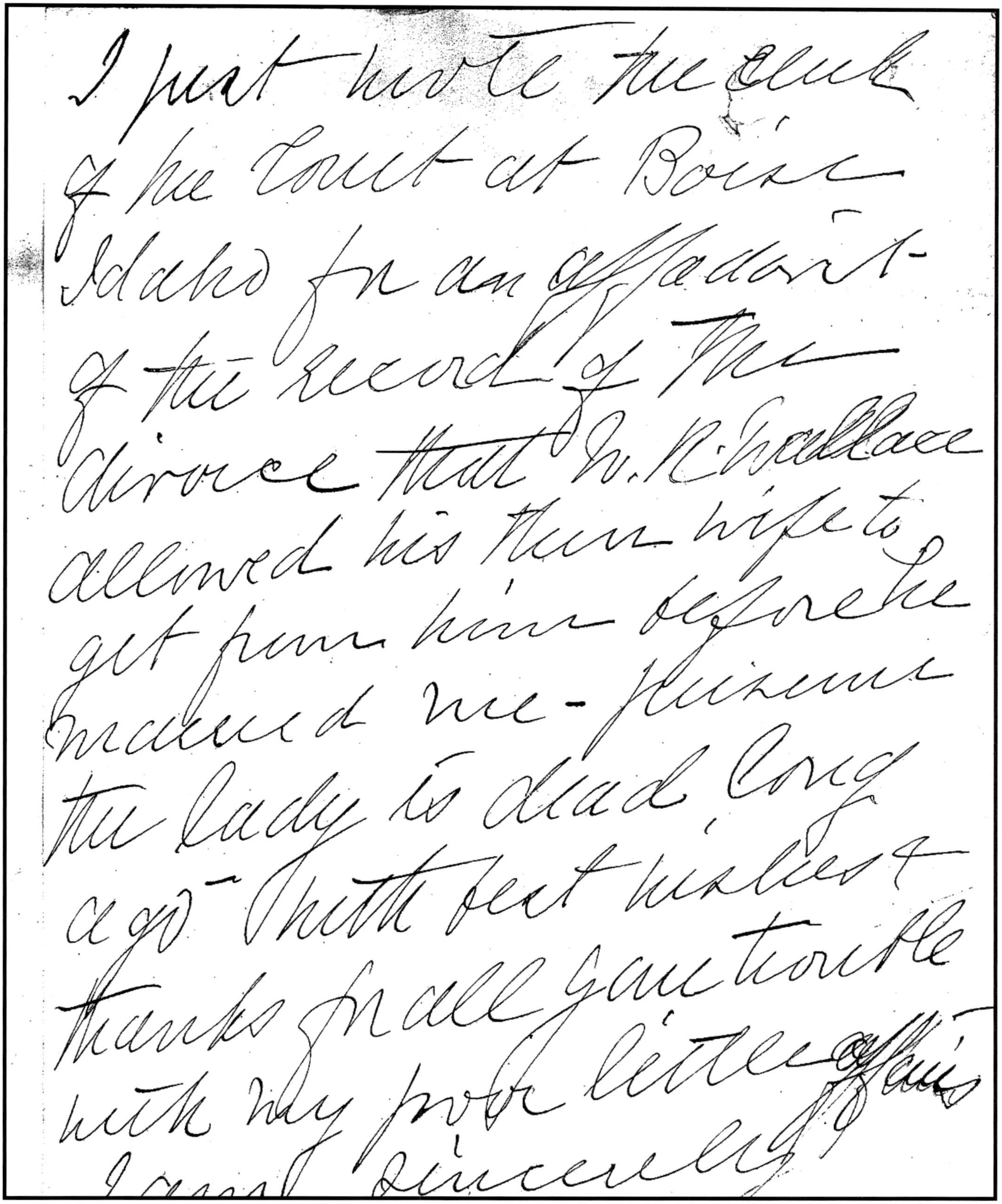

I just wrote the clerk
of the Court at Boise
Idaho for an affadavit
of the record of the
divorce that W. R. Wallace
allowed his then wife to
get from him before he
married me - presuming
the lady is dead long
ago - With best wishes &
thanks for all your trouble
with my poor little affairs
I am sincerely

The portion of the second page of a letter written by Annie Wallace to the "Widow Dept." regarding her claim for a widows pension. It states: "I just wrote the clerk of the court at Boise, Idaho **for an affidavit of the record of the divorce that W. R. Wallace allowed his then wife to get from him before he married me** - presuming the lady is dead long ago - with best wishes & thanks for all your trouble with my poor little affairs. I am sincerely A. E. Wallace"

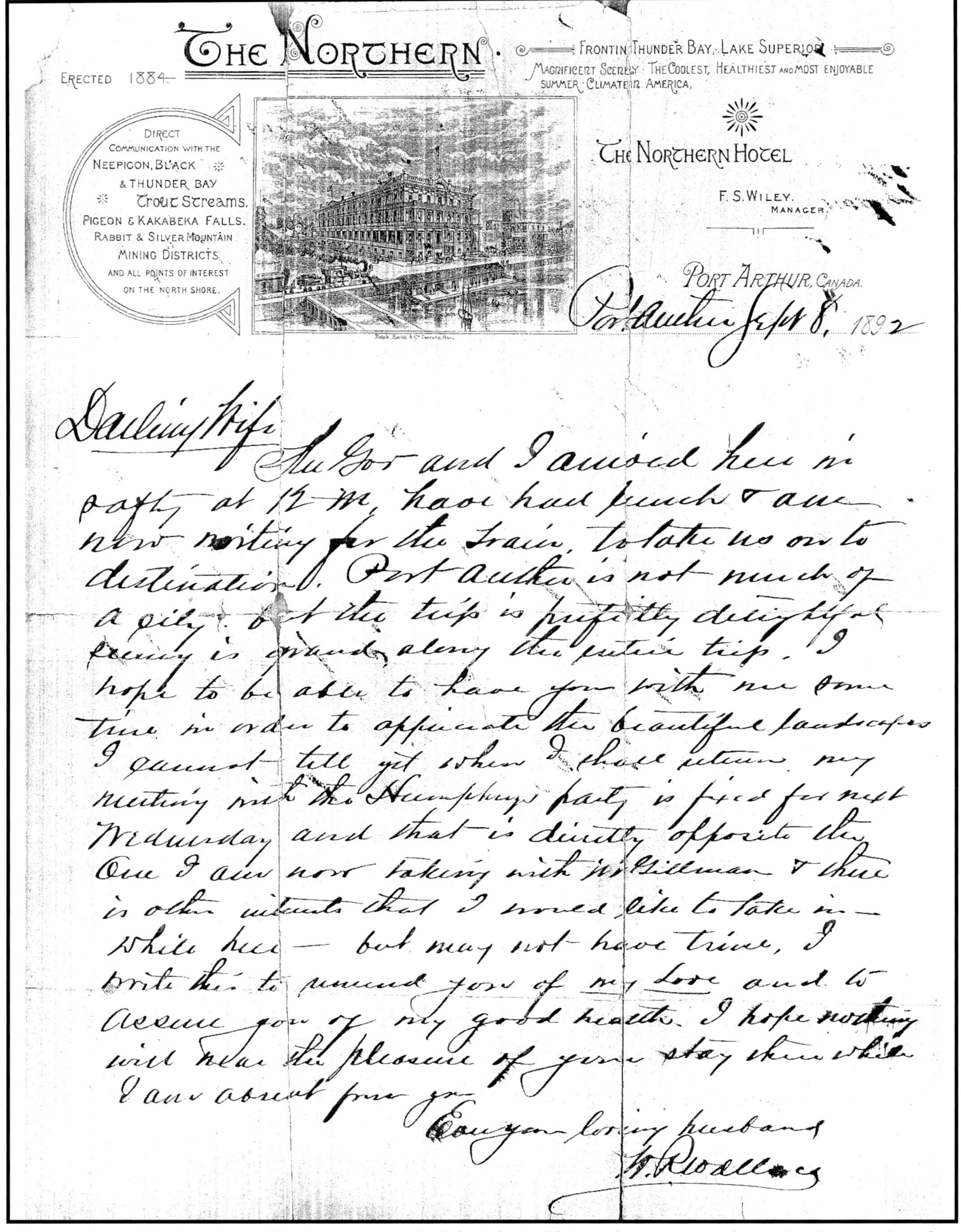

THE NORTHERN

ERECTED 1884

FRONT IN THUNDER BAY, LAKE SUPERIOR.
MAGNIFICENT SCENERY. THE COOLEST, HEALTHIEST AND MOST ENJOYABLE SUMMER CLIMATE IN AMERICA.

DIRECT COMMUNICATION WITH THE NEEPICON, BLACK & THUNDER BAY TROUT STREAMS. PIGEON & KAKABEKA FALLS. RABBIT & SILVER MOUNTAIN MINING DISTRICTS AND ALL POINTS OF INTEREST ON THE NORTH SHORE.

THE NORTHERN HOTEL
F. S. WILEY, MANAGER.

PORT ARTHUR, CANADA.

Port Arthur Sept 8, 1892

Darling Wife

The Gov and I arrived here in safety at 12 M, have had lunch & are now waiting for the train, to take us on to destination. Port Arthur is not much of a city. but the trip is perfectly delightful, scenery is grand along the entire trip. I hope to be able to have you with me some time in order to appreciate the beautiful landscapes. I cannot tell yet when I shall return. my meeting with the Humphreys party is fixed for next Wednesday and that is directly opposite the one I am now taking with Mr Gillman & there is other interests that I would like to take in while here — but may not have time, I write this to remind you of my Love and to assure you of my good health. I hope nothing will mar the pleasure of your stay there while I am absent from you

Ever your loving husband

W. R. Wallace

A letter from William Ross Wallace to Annie (his third wife), written approximately six months after they married. *(From the records of the National Archives)*

Lew Wallace (Unproven, but claimed cousin)

In William Ross Wallace's obituaries, of which there were many, most state: "He was the cousin of General Lew Wallace." This is very possible, as during my research, I have found William Wallace to have high integrity and have found his word to be good. **However, I was unable to substantiate that they were, in fact, cousins. One reason for this could be gleaned from a statement by Lew Wallace in some of his writings that <u>"Mine were folk who cared little for ancestry."</u> A favorable argument, however, is that William Ross Wallace's father's first name was Lewis and his mother's first name was May. Oftentimes relatives name their sons or daughters after their own close relatives. In my case, my dad named me after my Uncle Anthony (Tony).**

The following time-line is from the General Lew Wallace Study and Museum in Crawfordsville, Indiana:

• 1827 – Lew Wallace is born in Brookville, Indiana, on April 10 to David and Esther Wallace

• 1832 – The Wallace family moves to Covington, Indiana

• 1833 – Esther Wallace dies of consumption

• 1836 – David Wallace marries Zerelda Sanders

• 1837 – David Wallace takes office as the 6th Governor of Indiana; family moves to Indianapolis

• 1843 – Lew begins working as a copyist for the Marion County Clerk and begins studying the law at night

• 1846 – Lew serves as a 2nd Lieutenant of the 1st Indiana Volunteers in Mexican-American War, but sees little action

• 1849 – Lew passes the bar exam

• 1850 – Lew sets up law practice in Covington, Indiana

A Civil War portrait of Major General Lew Wallace, circa 1864. *(Public domain)*

• 1852 – Lew marries Susan Elston, daughter of Isaac Elston, on May 2

• 1853 – Henry Lane Wallace is born in Covington and family moves to Crawfordsville

• 1856 – Lew is elected to Indiana State Senate; Lew organizes the Montgomery Guards, a militia unit that gains him statewide notoriety

• 1861 – Lew is appointed Adjutant General of Indiana to begin organizing Indiana's soldiers; Lew becomes Colonel of the 11th Indiana Volunteer Infantry

• 1862 – Lew promoted to Major General and commands troops in Tennessee battles of Forts Donelson and Henry and Shiloh; Lew organizes a successful defense of Cincinnati, Ohio, in September

• 1864 – Lew is appointed Commander of the 8th Army Corps at Baltimore by President Lincoln; Lew leads troops in the Battle of Monocacy, which saves Washington, D.C., from Confederate assault

• 1865 – Lew leads secret mission in Mexico to stop flow of goods into the Confederacy; Lew is appointed second in command of the Lincoln assassination trial; Lew is appointed President of the Court for the trial of Henry Wirz, the commander of Andersonville prison

• 1866 – Lew returns to Mexico and supplies the Juaristas with arms in their successful attempt to overthrow Maximillian

• 1868 – Lew returns to Crawfordsville and builds a two-story Victorian home on land originally owned by Susan's father

• 1870 – Lew unsuccessfully runs for Congress

• 1873 – Lew's first novel, The Fair God, is published

• 1878 – Lew is appointed to serve as Governor of the New Mexico Territory by President Hayes

• 1880 – Lew's second novel, Ben-Hur, is published on November 12

• 1881 – Lew is appointed United States Minister to Turkey by President Garfield; Lew becomes good friends with Sultan Abdul Hamid II

• 1885 – Lew returns to Crawfordsville

• 1893 – Lew's third novel, The Prince of India, is published

• 1895 – Construction of Lew's study begins in Crawfordsville

• 1899 – Ben-Hur adapted into a stage play and opens at the Broadway Theater in New York City

• 1905 – Lew dies on February 15 in Crawfordsville

• 1906 – Lew's self-titled autobiography published with help from Susan and Mary Hannah Krout

• 1907 – Susan dies on October 1 in Crawfordsville

• 1910 – A marble statue of Lew placed in Statuary Hall at the United States Capitol Building in Washington. A copy of the statue is placed on the Study grounds in Crawfordsville

• 1925 – First full length motion picture of Ben-Hur released, starring Ramon Navarro

• 1959 – Remake of Ben-Hur starring Charlton Heston breaks box office records and garners a record eleven Academy Awards.

• 2015 – Filming begins for third full-length, live-action version of Ben-Hur, starring Jack Huston in the lead role.

Lew Wallace as a young man. *(Public domain)*

Descendants of William R. and Sarah (Slade) Wallace

After struggling through the challenge of piecing together the information on William Ross Wallace's three wives, which would have been nearly impossible without the help our friend Karen Curran, I thought a family history of Wallace's descendants would be an easier task. That could not have been farther from the true!

Karen was able to put me in touch with one known blood relative, Carla Elaine (Tulles) Heath, a great-great-granddaughter of William and Sarah Wallace. She was helpful with her direct family line, but also found the rest of the family history quite confusing. There were a lot of marriages, divorces, and remarriages within the family, which certainly complicated the issue. Karen, Carla, and my wife, Suzanne, compiled as much information as they could find, which I have recorded in this section. Thanks to Carla's involvement, there are photos of some of William and Sarah Wallace's descendants. The following narrative focuses on their two children, son Oscar B. Wallace and daughter Emma E. Wallace, and their direct descendants.

Son Oscar B. Wallace (1858–1939) and Family

Oscar Wallace was born April 3, 1858, in New York to William Ross and Sarah Slade Wallace. At the age of two, Oscar's family moved to Ohio. In 1870, they were living in Pennsylvania, and in 1880, Centerville, Appanoose County, Iowa. They no doubt were following the mining booms.

On December 8, 1881, at the age of 23, Oscar married Laura Bell Scott in Centerville. Laura B. Scott was born in Knoxville, Marion County, Iowa, around 1862 to William M. Scott and Sarah Rebecca (Neuse) Scott. William Scott was a prominent physician in Centerville, where the Scotts had moved at the close of the Civil War, during which Dr. Scott had served as a physician.

In 1886, Oscar, Laura, and their two-year-old son, Walter Scott Wallace, left Centerville to join Oscar's father, William R. Wallace, at Placer Center (Wallace), Idaho. Oscar would later become the town's mayor (1894–1895). Their second child, Carrie Saffelle "Caddie" Wallace, joined the family in November 1888, the first white child born in Wallace. (Saffelle was Laura's grandmother's maiden name.) Ruth also was born while the family was living in Wallace. According to her headstone, she was born on April 12, 1893, and was only ten years old at the time of her death on October 2, 1902. She died at Sacred Heart Hospital from peritonitis and was buried in the family plot at Spokane's Greenwood Memorial Cemetery.

Oscar B. Wallace. *(Courtesy Butch Jacobson)*

By the time of the 1900 census, the family had relocated to Spokane and Oscar was listed as a silver miner. On July 29, 1905, a daughter Laura was born in Spokane. By all appearances, she did not live long, though no death record or place of burial could be located. Daughter Helen K. was born the following year. An article about Laura Scott Wallace after her death stated she was preceded in death by two sons (Walter and Frederick William) and two daughters (Carrie and Ruth – but, strangely, no mention of Laura). We found nothing during our research process that mentioned a son Frederick William, and the only other reference we found to him came from

Carrie's granddaughter Carla Tulles Heath, alluding to him being buried in Greenwood (though a record search there did not produce information on him).

In 1910, Oscar is shown on the census in Spokane with his family. His occupation is "miner." However, he is also on the Wallace 1910 census, living on Bank Street, and his occupation was "manager of lead mine" (probably the Callahan, as he was a third owner and the manager).

Granddaughter Carrie (Wallace) Tulles (1884–1911) and Her Descendants

Sadly, in 1911, daughter Carrie died a tragic death, the result of being severely burned from the explosion of a five gallon can of alcohol. She lingered for 10 days, and died on August 25, 1911. She was survived by her husband of less than two years, Russel Parker Tulles, who opened a drugstore in Kennewick, Wash., and their nine-month-old son, Oscar Wheeler Tulles (Carla Heath's father). Oscar was born in Kennewick on December 1, 1910. Carrie was only 22 years old at the time of her death and was the third daughter of Oscar and Laura Wallace's to die within a nine-year period. She is buried in the Greenwood Cemetery with her sister Ruth.

Carla said it appears the Tulles family (mostly Russell's sister Stella) helped raise little Oscar until Russell remarried when Oscar was about seven. Oscar then moved to Denver with him and his new wife. Oscar married Roberta Scott on July 21, 1934. Their daughter, Carla Elaine Tulles, was born in Texas (where she still lives) on August 29, 1943. She married Patrick Roy Heath on June 18, 1965. They are the parents of two daughters, Jennifer Rebecca Heath (married to Thomas Ross Porter) and Laura Rachel Heath (married to Thomas Richard Hudgings). The Hudgings have two sons, Benjamin Walter Hudgings and Oscar William "Gus" Hudgings. Carla's father, Oscar, died September 30, 1986, in Bexar County, Texas. Her mother, Roberta, died in San Antonio, Bexar County, Texas, in July 1984.

Carrie "Caddie" (Wallace) and Russell Parker Tulles. They married January 12, 1909, and lived in Kennewick, Wash., where Russell opened a drugstore. They had one child, Oscar Wheeler Tulles, who was only nine months old when Carrie died.

Oscar Wheeler Tulles, circa 1915, great-grandson of William and Sarah Wallace. From the sounds of a letter written to him in 1931 by his grandfather, O. B. Wallace, they were nearly inseparable when he was a little boy. After the death of young Oscar's mother, Carrie, O. B. tried to talk his son-in-law into letting the little boy stay with him. Oscar went into the mines with O. B. and had his own pick, which he is holding in the photo. The photo was at one of O. B.'s mines.

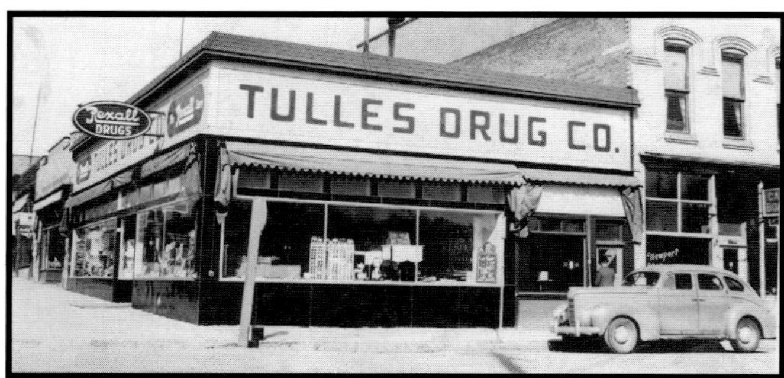

Tulles Drug Company in Newport, Wash., circa 1940. Carrie's husband, Russell, his father Dr. J. W. Tulles and brother Arthur Tulles were partners in this company for a period of time. Russell later decided to become a dentist and sold his interest to Arthur. *(All photos this page courtesy Carla Tulles Heath)*

Left: Laura R. (Heath) and Thomas R. Hudgings on their wedding day, May 9, 1998. Right: The Heath family in July 2015. The adults, from left, are Patrick Heath, Carla Tulles Heath, Laura Heath Hudgings, M.D., Jennifer Heath Porter, Tom Hudgings, and Tom Porter. The boys in front are Carla and Patrick Heath's grandsons, Benjamin Walter and Oscar William "Gus" Hudgings. *(Courtesy Carla Tulles Heath)*

Granddaughter Helen Wallace and Her Mother, Laura (Scott) Wallace, Move to Iowa

By 1915, Laura and their daughter, Helen, then about 10 years old, were living with Laura's parents in Centerville, Iowa. The 1925 census lists Laura Wallace as still living with her mother, Sarah Rebecca (names sometimes reversed) Scott and daughter, Helen, in Centerville, Iowa. The census lists her marital status as divorced (the 1930 census indicates widowed, even though Oscar was still alive and had remarried). Laura and Helen continued living with Sarah until her death in May 1935 (William Scotthad died earlier), but they remained in Centerville. The 1940 census shows them at the same residence. After their respective deaths, Laura in 1945 and Helen in 1951, they were buried in the Oakland Cemetery at Centerville, where they share a headstone. It does not appear that Helen ever married.

Grandson Walter Scott Wallace (1884–1920)

Walter Scott was born in 1884 (dates conflict) in Centerville, Appanoose County, Iowa, to Oscar B. Wallace and Laura Bell Scott. As mentioned earlier, in 1886, his family moved to Wallace, Idaho, where he was said to be the first white child. Walter became a physician, having received his medical education at the College of Physician & Surgeons in Louisville, Kentucky. He practiced his profession in Tekoa and Spokane, Wash., before settling in Newport, Wash., in 1911. He was a well respected member of the community.

Walter was married three times and had one child, Mercedes Bales, born in Spokane on June 20, 1911. She was the daughter of Walter's second wife, Verna Starbird. According to Carla Heath, Mercedes had one child, daughter Brenda, who was killed in a car accident when she was probably in her 30s.

On September 24, 1916, Walter joined the Royal Army Medical Corps. He held the rank of captain and saw service on the Saloniki and Italian fronts during World War I. He died on December 19, 1920, while hunting around Priest River, Idaho, The cause was determined to be heart failure. He is buried in the Wallace family plot at Spokane's Greenwood Cemetery, along with his sisters Ruth Wallace and Carrie (Wallace) Tulles.

Final Days of Oscar B. Wallace

Oscar and Laura Wallace both outlived all but one of the children (Helen), who may not have had much of a relationship with her father after her parents separated. By the time of the 1930 Los Angeles census, Oscar was married to Nelle. She was 45 years old and Oscar, listed as a poultry farmer, 73. According to some family history, Oscar may have had another marriage between the time he was married to Laura and when he married Nelle. There was even mention that he was married to two women at the same time. According to Joan Sellers, who claims she is Colonel Wallace's granddaughter (would technically be step-granddaughter), Oscar was married to her mother, Helen (last name unknown). We have nothing to substantiate any of these claims.

According to a letter Oscar wrote to his namesake grandson, Oscar Tulles, in 1931, he had traveled all over the country and into Canada pursuing mining ventures, just like his father before him. He said, like thousands, the Depression rendered most of his properties worthless. From the letter, it sounded as though he and his grandson had been very close, but lost contact when little Oscar returned to live with his father and they moved to Denver. It was 15 years later that his grandson located his whereabouts and wrote a letter, the one to which Oscar B. responded.

Oscar B. Wallace died from heart complications on September 20, 1939, at the Los Angeles County General Hospital. His death certificate listed his occupation as a "mining promoter" for 40 years, but that he last worked in that occupation was 1923. No obituary could be found, but an Official Death List in the *Los Angeles Times*, on the date of his death listed his and age of 81. He was buried in the Grand View Cemetery.

Daughter Emma E. Wallace (1860-1953)

Emma E. Wallace was born in January 1861, in Alliance, Ohio, to William Ross and Sarah Slade Wallace.

In the 1880 census, Emma was living with her parents in Centerville, Appanoose County, Iowa. She has the last name Lohnridian, but was listed as divorced. She had a two-year-old son, Robert Fay Lohnridian. A few months later, in December, Emma married Frederick A. Neuse, who was the brother of Sarah R. Scott, the mother-in-law of her brother, Oscar B. Wallace. Emma and Fred had a son, Roy, who was born in 1883 and died in 1900, and a daughter, Florence, who was born in 1886, the same year her father died. Emma's firstborn son was listed as Robert Neuse, the name he carried for the rest of his life, so it appears he was adopted by Fred Neuse.

Emma married Thomas Goss in Illinois in March 1891, and together they had a son named Edwin Earl Goss, born in 1892 in Centerville, Iowa. Emma and her four children were still living in Centerville at the time of the 1900 census, but she was listed as divorced and head of the household. However, the 1910 census lists years married as what appears to be 21 (the entry is faded, but it would be consistent with when she married Thomas Goss). She was then Denver and was said to be 38 years old, an obvious error.

By 1920, Emma was living with her son Robert, in Tucson, Arizona, and was listed with the same last name as her son, Neuse. Her marital status showed widowed, though Thomas Goss did not die until 1926. However, his death record named another woman as his spouse, so Emma was probably divorced from him. She died in Phoenix, Arizona at the Butler Rest Home on April 2, 1953. She is buried in Greenwood Memory Lawn Cemetery in Phoenix, under the name Emma Neuse. On her death certificate, there was no known information about her, which means she did not have family nearby to contact. Robert, who had never married, died in 1951.

Nothing could be found about Florence after 1905, when she was listed on census for Mapleton, Blue Earth County, Iowa, along with her mother and younger brother, Earl Goss. Emma's mother, Sarah, had died the previous year, and her obituary stated her daughter was with her at the time of her death.

Earl Goss, Emma's youngest child, was living with her in Denver in 1910. Around 1914, he married Maybelle Wilcox. Denver remained their home for many years, but they eventually moved to California, where Earl died in December 1969. They had a son Earl Jr., who was born around 1915 and died in 1987 in Denver, and a daughter, born around 1917, whose name was listed as Ruth on the 1920 census and Yvonne in 1930 (the ages suggest they were the same child). Earl had two children, David Goss and Daniel Goss.

The Gold Rush Brought Wallace to the Coeur d'Alenes

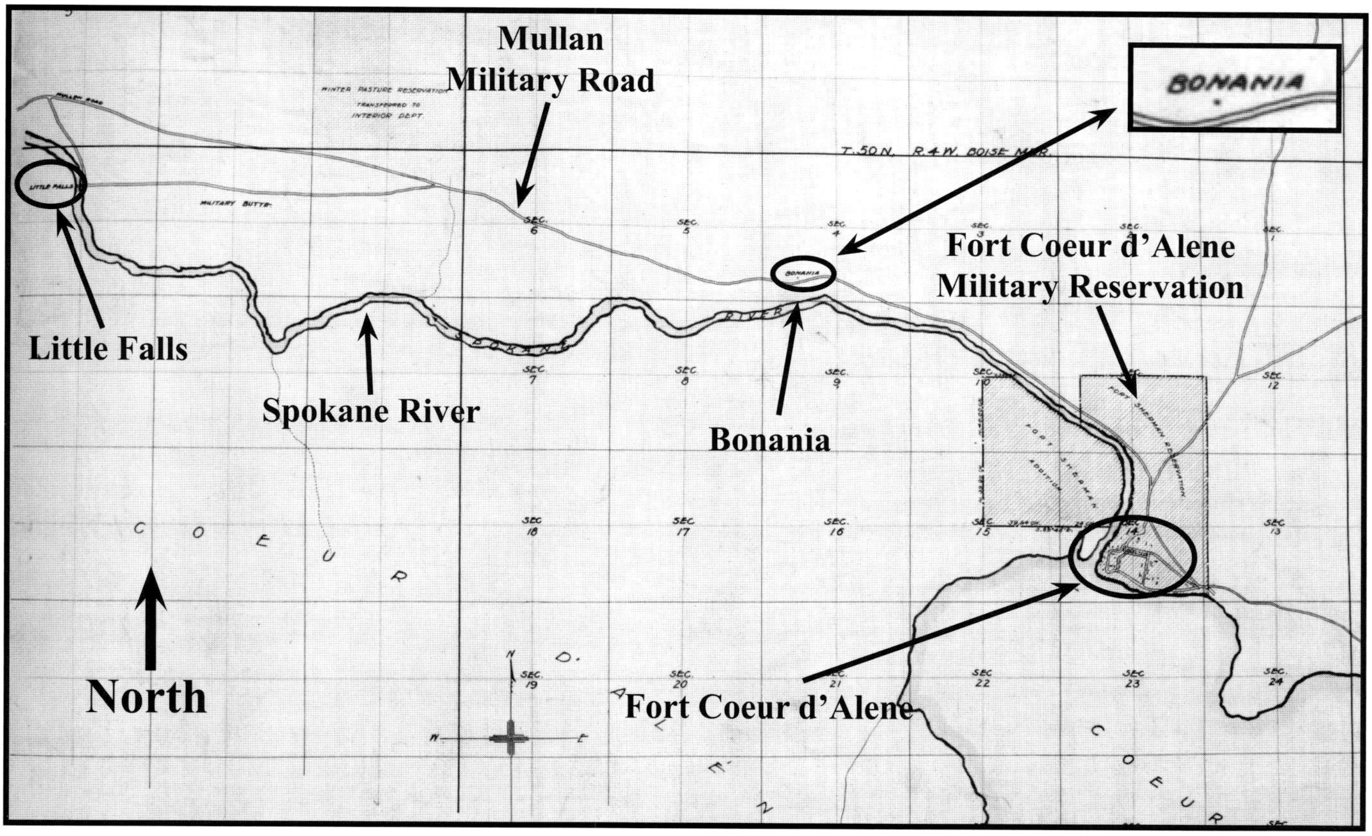

Original 1878 military map of Fort Coeur d'Alene and surrounding area. *(Courtesy Museum of North Idaho)*

Prior to the discovery on Prichard Creek that ignited the 1883 gold rush, Andrew Prichard and Tom Irwin were prospecting along the Mullan Road. They each had made some discoveries near present-day Osburn, Idaho, and were busy working their claims and building shelters. However, during the winter of 1879-80, they lodged at Bonania City, near Fort Coeur d'Alene. It may have been because the winter weather was not as severe as where they had been prospecting or because there was better lodging, but it appears it was only a temporary arrangement. They returned to their claims along the South Fork of the Coeur d'Alene River. Soon, venturing over the divide, Prichard and his party discovered gold in more ignificant quantities. Though initially intending to share the news with a select few, the word leaked out and the rush was on. Among the first arrivals was Colonel William R. Wallace.

Bonania City

Although the original intent of Congress when it established the census was to distribute taxes among the states based on population, an official census is the most dependable and accurate measure of a community's development and growth.

At the time of the 1879-80 census, Kootenai County, much larger geographically than at present, was included in the 10th United States Census. Fortunately, for historical purposes, it couldn't have happened at a more auspicious time. With the completion of the Northern Pacific Railroad only three years away and the projected commerce it would yield, a small but steady influx of people began arriving in the area just prior to the establishment of Fort Coeur d'Alene on April 16, 1878. (Its name was changed to Fort Sherman on April 12, 1887, by government order 30, A.G.O. 1887.)

During that time period, strained race relations resulting from the war against Chief Joseph and the Nez Perce Indians made interactions between the Indians and all newcomers tenuous at best. Consequently, military forts were still considered a major necessity for the peaceful settling of the West, and the presence of such a fort in Kootenai County would prove to be a great stabilizing force for that area and an attraction for development.

The officially recorded settlement of Kootenai County began with the publication and release of the 1878 military map of Fort Coeur d'Alene and the surrounding area, followed by the records generated by the United States census takers in 1879-80, acting in sync with the planned establishment of the new military fort. These two written documents made record of the first official settlements and occupancy of that area. According to the 1879-80 U.S. census, along with a scattering of people throughout Kootenai County, there were just a few recognizable communities, including the Military Post Coeur d'Alene; Little Falls, which later became Post Falls (population 22, mostly lumbermen); Bonania City situated approximately three miles west of the military post, and Rathdrum. Once the gold rush started in 1883, this situation would change radically. The settlement at Fort Coeur d'Alene quickly grew into the town of Coeur d'Alene. Many early western towns had "city" added to the name, although the only criterion appeared to be a concentrated cluster of dwellings with some type of public services.

Most important, for the purpose of clearly establishing Kootenai County's early history, Bonania City was already established as an existing community prior to any other communities in the county. In fact, it was established to the extent that the post commander and two of his subordinate commanding officers chose to live in pre-existing houses at that location rather than in tents at the fort site during its construction.

There are contradictions regarding the spelling of Bonania City. According to the handwritten 1879-80 census, it was referred to as Bonanzy City. Later, when that census record was typed up, the town was listed as "Bonanza." On a military map dated August 25, 1879, which laid out the proposed Fort Coeur d'Alene military reservation, the name "Bonania" was used. It is the first known map of that area and, as an official military document, the spelling of the name is likely to be the most accurate of the various options.

Bonania City, as shown on the military map, appears to have been located at or near present-day Huetter. At the time of the census, the town was comprised of 15 dwellings and two commercial enterprises offering services.

In addition to this map and the information contained in Kootenai County's first United States census (1879-80), there are two other corroborating documents that confirm the existence of Bonania City. One of these documents is a manuscript in the Museum of North Idaho written in the early 1900s by P. W. Johnson from an interview with Coeur d'Alene's first boat builder, Peter Sorensen. Sorensen was commissioned by Fort Coeur d'Alene's commanding officer to build the first steamboat on Lake Coeur d'Alene, the Amelia Wheaton – named after the post commander's oldest daughter. In this interview, Sorensen gave the following description of his trip into Coeur d'Alene, which, although filled with spelling errors, is of great historical value:

The last stop the stage made before comeing to Fort Coeur d'Alene, was at Banansa City. Here the horses were watered and given their Oats, as was always don, and we, the Passengers to the Dineing room. In the dineing room they had tacked som Chese-cloth to the Loges overhead, and we all know how the flies work, and leave black spots, resembling bullet holes in the Sealing. Seeing this Sorensen made the statement, that we were now getting into real country. Just look at the bullet holes in the Sealing. The Tenderfoot went straight up, and remaned so as long as they were in there. This city contained a log house, and a Barn also built of Logs and was situated across the now Highway, west of Ohio Match Saw Mill [just west of Coeur d'Alene on the Spokane River]. At this point there are a Bar across the River, where it was crossed at low water.

The second source is *History of North Idaho* (1903):

. . . With the advent of the railroad came the desire for county organization. In July 1881, M. D. Wright and George B. Wonnacott called a meeting for all the settlers known to be in the county for the purpose of discussing the question. Two or three meetings were held and much canvassing was done before the number of required petitioners could be obtained. Two of those meetings were held at Mr. Wonnacott's store two miles west of Fort Coeur d'Alene and the third meeting was held at Rathdrum, then called Westwood, where the organization was finally completed in October 1881. . . .

The George Wonnacott referenced above was listed in Bonania City in the 1879-80 census (see following page). He was born in Canada in 1840, and was a single, white male whose occupation was listed as a "dealer in general merchandise." From this information it can be concluded that Bonania City was a rest stop between Spokane Bridge and Fort Coeur d'Alene, with at least three commercial establishments: a general store, a business establishment where meals could have been purchased, and some type of commercial stable facility. There also may have been additional enterprises, as other occupations listed included schoolteacher, barkeeper, trader, cooper (barrel maker), two shingle makers, and numerous farmers and laborers.

Cœur d'Alene Mines!

The Gold Excitement at Fever Heat!

New Developments with Rich Results!

A Perfect Stampede into the Diggings!

THE RICHEST PLACER MINES ON THE COAST!

Gold – the reason Colonel Wallace came to the Coeur d'Alenes

A sample of the type of news clippings about the gold discovery that appeared in the September 1, 1883, and September 22, 1883, issues of the weekly *Spokane Falls Review*. These were the first of a long-running and wordy series of newspaper articles covering the gold strike in the Coeur d'Alenes. As a result of the newspaper coverage, and the fact that the *Spokane Falls Review* was a member of the Associated Press, the word of the gold rush was fairly well spread prior to the NPRR's advertisement campaign, which did not begin until February 1, 1884. When the quantity of the placer gold ran out, the newspapers continued with news of the world-famous lead-silver strikes.

Who's Who in Bonania City, Idaho, in 1879-80

According to the 1879-80 United States Census for Kootenai County, its present population was a total of 508, which included 289 soldiers and some civilians living at Fort Coeur d'Alene. At nearby Bonania City, there were 50 people occupying the 15 structures, including the commander of Fort Coeur d'Alene and two of his fellow officers. As a matter of historical significance, the first prospectors of the 1883 gold rush to the Coeur d'Alenes – Andrew Prichard, Thomas Irwin, and William Gerrard – were presently at Bonania, along with four army officers, their families and servants. Andrew Prichard and Thomas Irwin were sharing Dwelling #43 and were listed as placer miners. Although it's thinly veiled historical fiction, in May Hutton's *The Coeur d'Alenes or a Tale of the Modern Inquisition in Idaho* (published 1900), she wrote that Tom Irwin prospected "near the military post on Lake Coeur d'Alene for a couple of years, . . ." The "couple of years" part is inaccurate, but it is natural to assume they prospecting the area near Fort Coeur d'Alene while living in Bonania.

The Bonania City 1879-80 Census

Dwelling #32
Wonnacott, George, 49, ... Dealer in general merchandise
O'Connor, Michael, age 25 ... Bar keeper

Dwelling #33
Lemley, Christopher, age 26 ... Wood chopper
Carnes, George, age 25 ... Farm labor
Hale, Thomas, age 24 ... Farm labor
Theodore, Adolphe, age 34 ...Trader

Dwelling # 35
Cohorn, Robert, age 43 ... Farmer
Cohern, Lela, age 32 ... Wife, keeping house
Cohorn, Robert, age 13 ... son, at home
Cohorn, Nettie, age 12 ... daughter, at home
Cohorn, Altie, age 10 ... daughter, at home

Dwelling # 36
VanOlstein, Andrew, age 39 ... lumberman

Dwelling # 37
Johnson, William, age 38 ... farmer

Dwelling # 38
Clapp, Alvaro, age 31 ... farmer

Dwelling #39
Thrasher, George, age 22 ... farm labor
Tackaberry, Bery, age 31 ... school teacher

Dwelling # 40
Hayden, Mathew, age 41 ... farmer
Berry, George, age 32 ... farm labor
Stanley, Edward, age 26 ... farm labor

Dwelling # 41
Paschal, James, age 28 ... "cooper" (barrel maker)

Dwelling # 42
Richey, Alfred, age 41 ... shingle maker
Woodfirs, James, age 24 ... shingle maker
Stewart, William ... shingle maker

Dwelling # 43 (The two men were listed as partners)
Prichard, Andrew, age 45 ... placer miner
Irwin, Thomas, age 49 ... pacer miner

Dwelling # 44
Gore, William, age 37 ... laborer
Hitch, John, age 33 ... laborer
Gerrard, William, age 41 ... farm laborer

Military dwelling # 45
Commander of Fort Coeur d'Alene (Sherman)
Wheaton, General Frank, age 47 ... soldier, general
Wheaton, Maria, age 34 ... wife, keeping house
Wheaton, Amelia, age 13... daughter, attending school
Wheaton, Octavia, age seven months ... daughter, at home
Schrutchins, Barbara, age 56 ... servant, nurse
Schrutchins, Charles, age 14 ... servant, errand boy
Ching, Ah, age 23 ... servant, cook

Military dwelling # 46
Mills, William, age 43 ... soldier, captain
Mills, Jennie, age 33 ... wife, keeping house
Mills, Anna, age 4 ... daughter, at home
Mills, William, age 2 ... son, at home
Loo Teung, age 38 ... servant, cook

Military dwelling # 47
Keller, Charles, age 37 ... soldier, captain
Keller, Mary, age 32 ... wife, keeping house
Keller, Annita, age 2 ... daughter, at home
Keller, Mary, age six months ... daughter, at home
Robinson, Jane, age 17 ... servant, nurse
Lee Sing, age 28 ... servant, cook

Military dwelling # 48
Clark, Sidney, age 46 ... soldier, captain
Clark, Ellen, age 36 ... wife, keeping house
Clark, Anna, age 13 ... daughter, attending school
Brinkman, Margaret, age 18 ... servant, nurse

The Mullan Military Road and Surroundings Between Kellogg and Wallace at the Time of the 1883 Gold Rush

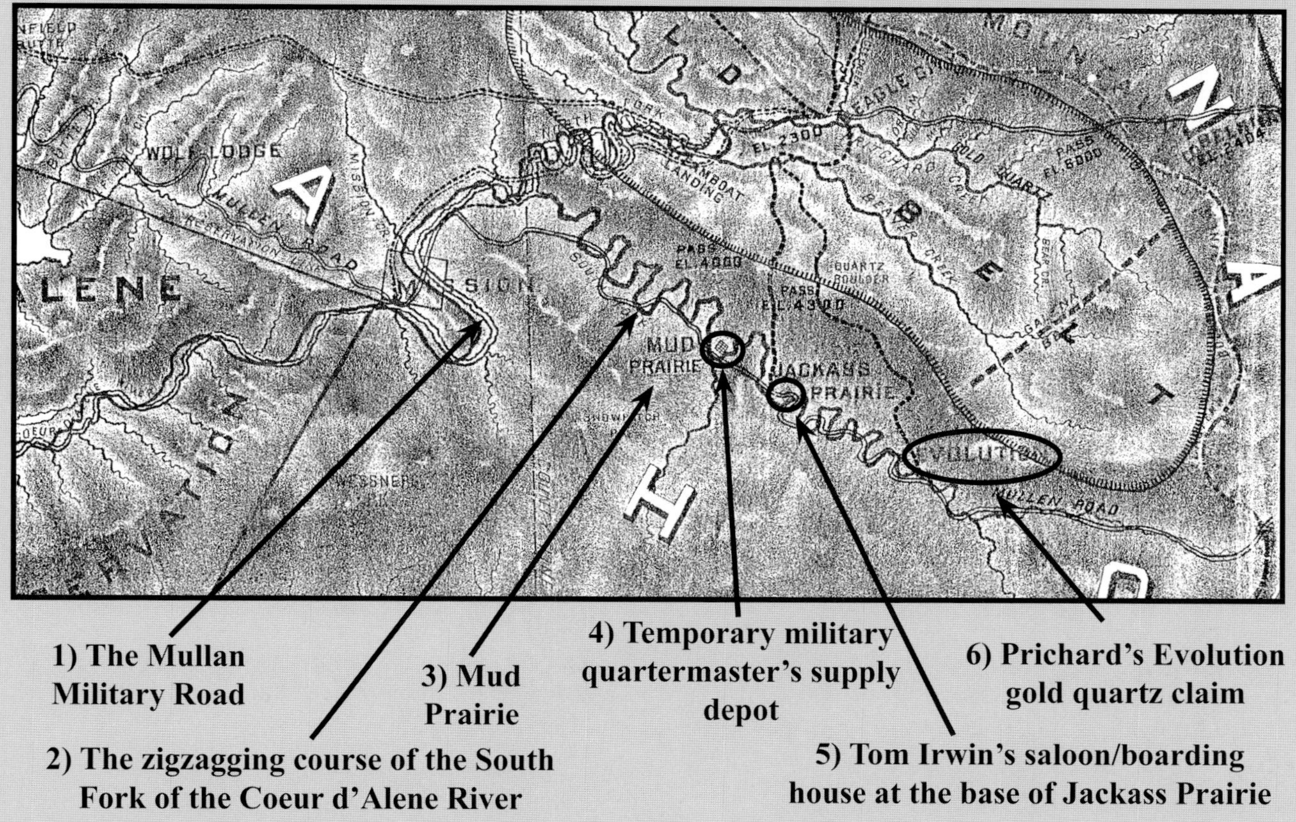

This map, especially significant because it shows six major features that existed prior to the major development of the area, is the earliest depiction of that area as it appeared in 1883-84. It was compiled on January 25, 1884, by F. L. Miller, Civil Engineer, and titled: *Map of the Coeur d'Alene Mines and Vicinity, Idaho Territory*:

1) The Mullan Military Road is depicted as a double black line, consistently either skirting or fording the South Fork of the Coeur d'Alene River for a distance of approximately 20 miles.

2) In an article that appeared in the *Oregonian* on July 3, 1897, Adam Aulbach describes this stretch of the Mullan Military Road, which today would be the region between Kellogg and Wallace: "The country was very inhospitable at the time. The valley of the south fork of the Coeur d'Alene river was densely covered with timber, and the Mullan road which made its sinuous way through it was scarcely perceptible, **while every half mile or so the stream had to be forded**. The mountains all about the region were also heavily covered with timber, while there were no habitations or supplies within a hundred miles or more."

3 & 4) During the construction of the Mullan Road one of Lt. Mullan's men was Charles Schafft. Many years later, when Schafft began working for the first Fort Benton newspaper, the *Benton Record*, he wrote a series of "Literary Contributions." Schafft's first contribution was a reminiscence of his time with Lieutenant Mullan during the construction of the Mullan Road. A portion of Schafft's account, which was published in the January 2, 1880, *Benton Record Weekly*, described the Mullan Road between what is now Kellogg and Wallace.

In his narrative, Schafft described both the military depot shown on this map and Mud Prairie:

> *Mud Prairie, eleven miles above the Mission, was fixed upon as a depot camp. This prairie, naturally a swamp, was made more so by the previous heavy rains, and had to be partly bridged to get the wagons to its upper end. An examination of the surrounding hills found them full of springs and impractical for grading. We were now at the main barrier of the entire road, and it was a serious one. The pass on both sides was obstructed by an almost impenetrable heavy growth of pine, cedar, tamarack and fir, long since thinned out by frequent fires occurring almost annually for the past twenty years. The mountains hugged the streams so closely that numerous crossings or time-consuming or laborious grades, were unavoidable. The timber on the line of the road had been set on fire, probably by Indians, and everything looked smoky, dismal and discouraging; but gloves had to be laid aside now, and working parties provided with eight or ten days' rations were pushed ahead to cut out, inch by inch as it were, the timber marked by the engineers, who were crawling through the undergrowth, unable to see more than a few feet before them. The road followed the bottom of the canyon, because it would have taken nearly a whole summer's work to grade the hills, even if that were practicable; as it was it took nearly the whole month of October to open a merely passable way for the wagons from Mud Prairie to the summit, a distance of only twenty-five miles, and the men were working hard from the earliest dawn till dusk. Mullan, who did not hesitate to put his shoulder to the wheel, was ever among them, to instill courage and hurry up the work. A fall or two of snow began to warn us of the approaching winter. While at Mud Prairie a Quartermaster's train brought out the winter supplies for the military escorts, which necessitated double tripping on part of our teams to the next depot at the foot of the mountains.*

5) This article by Newton H. Chittenden, titled *The Gold Fields of Coeur d'Alene*, appeared in the February 15, 1884, edition of the *New York Herald*:

> *JACKASS PRAIRIE OR HUNTERS HALL, Seven miles beyond was the next point reached of historic interest. This little opening easily cleared by a jack rabbit of average agility in a dozen leaps commemorates the meeting here on two different occasions by the Catholic fathers with a pioneer riding a jackass. Hunters Hall it should be called henceforth, for its cabin, the home of Tom Irvine [Irwin], . . .*
>
> *Between the Mission and Evolution we crossed the Coeur d'Alene river twelve times. In all but two places, fortunately, it was either frozen over or bridged by fallen trees. At one of these a ferry boat with an absent owner was pressed into service, and over the other I was safely carried on the back of a stalwart mail carrier, Joe Mulheur. At Evolution the labor of the Journey commenced in good earnest. Both the trails, Evolution and Trout Creek had been represented to me by old timers who had tried them as a "holy terror," and after a personal experience of sliding more rods down steep places on my corduroy pants, and turning more half somersaults backward and forward and plunging head foremost into more snow banks than on all of my previous travels, I can certify that for once they were very moderate in their statements.*

6) The Evolution is commonly known as the oldest claim in the Coeur d'Alenes, Andrew Prichard having first staked it in 1881 or 1882. The claim is situated on the north side of the South Fork of the Coeur d'Alene River, a little over a mile west of Osburn. Very little ore showing was ever found on this claim.

As Seen Three Years Ago, An Interesting Letter Written in the Darkest Hour of the History of the Coeur d'Alene "Coeur d'Alene Record"

The above was the headline for an article printed in the *Oregonian* on May 10, 1887. It was written by William Wallace in 1884 during a visit to Portland. In it, he described the massive lead and silver potential of the Coeur d'Alenes. Of significance is how clearly it illustrates the intelligence of William Wallace, the breadth of his mining knowledge, and his description of the area in 1884.

The *Oregonian* offered this introduction: "A short time ago the Record's editor came across the following interesting article written for the Oregonian nearly 3 years ago by Col. W. R. Wallace, who was then visiting Portland. It loses none of its interests through its age. Since first coming to Coeur d'Alene, Col. Wallace has been sanguine of its great future, and three years have well sufficed to prove that the faith which enabled him to weather the darkest period in its history was well-founded. The article is an interesting contribution to the newspaper literature of Coeur d'Alene and we printed it without alteration:"

All new departures, either in invention or discoveries of whatsoever kind, have opposition and criticism. The discovery of gold in California in 1848 made many a heart ache, and many a homeless wanderer. But it made almost an Empire of the Pacific coast and the fortunes. The invention of the Morse system of telegraph met with derision from statesmen to plowboy, but it enlightened the world by its influences.

The skepticism of mankind, and a conceit to have others think we know what in truth we do not, has led astray some of the world's beneficiaries, whether from nature's storehouse or the brain of man.

Of all mining camps, those of the Coeur d'Alene have, perhaps, been the most unfortunate as regards the tramps and "vags" who have thought fit to inflict their presence or their ideas on those who went there to work out the problem of pay on bedrock. Letters have been written and advice given enough to make misery for the million, and yet the earnest workers are well satisfied with the mines of Coeur d'Alene or of that belt of mineral extending from the North Fork across to the South Fork of the St. Joe River.

The writer left Portland eight months ago in a position and under circumstances to learn the true inwardness of the mining interest in these named localities; and has been there since without vacation, and has assisted to develop a section until lately unknown, even to the early prospectors of that region.

Leaving the beaten trail to Eagle, Prichard, Murray and other camps on the North Fork, we took the South Fork from Evolution. Moving our supplies from Kingston step-by-step, across the streams on ice, logs, or wading the surging waters, until the snow had gone sufficiently to enable us to begin our labors. Having information of silver ores being discovered on Beaver, we reasoned that between these points rich fields could be expected. About one mile above Evolution we made our first halt, and there drew our first conclusions; that some time in the past the South Fork had been a vast river, that the sedimentary formation was foreign to that location, that the boulders of quartz were brought there from the northeast; and finding this wash on the mountain ranges as well in the gulches confirmed this opinion.

Tracing this along the range up to the South Fork, we found above a certain line (nine miles) it ceased and came in from the north, with two forks of the South Fork of the Coeur d'Alene, one of which was named Nine Mile creek, the other Canyon creek, and up these we followed the sedimentary flow or wash to the line of mountains, in the Bitter Root range, near Thompson Falls; and during all our trip made careful examinations of the different stratus of native rock, but nowhere found the same as in the boulders of quartz that formed the wash.

Another fact we learned, that whenever on mountainside or valley an obstacle of any solid formation that could have formed an eddy in the old river-way, we could find gold, and on Nine Mile creek seemed the main channel for this deposit of quartz wash.

We located our placer claims, one of which we commenced working as soon as the snow would permit, on which we reached bedrock July 9 with a very satisfactory result, taking $17.90 from four feet square, our first test. Our sluice ditch is 400 feet long, 17 feet 6 inches deep at the head. This result satisfied us that no gold in any quantity will ever be found in place in the quartz veins of Coeur d"Alene, but that in the area of this river wash, which is from eight to twelve miles wide, and can be traced for thirty miles in length, the richest mines on bedrock can be expected and are already found, ever yet known. This is an established fact in Dream gulch, on Prichard creek and Beaver, besides the one mentioned above.

The conclusions are, then, that at some far back period of time the drainage of Clark's Fork of the Columbia flowed across where the sedimentary flow now lays; that ice and strong currents of water have brought the gold from the mother lode and ground the rocks and boulders by constant wear until the gold has been separated and deposited on another field – cleansed of quartz and ready for hardy hands at the expense of blows from pick and shovel, but none for those who expect to gather it without labor or expense.

Our explorations after the gold-bearing quartz led us up the range that separated the headwaters of the North Fork from those of the South Fork of the Coeur d'Alene, between Nine Mile and Canyon creek's, finding the country rock nearly all slate until about three miles north of the Mullan road. Here a stratum of granite of a mile in width was found, cutting across the ranges from the head of Beaver creek to the St. Joe river, a distance of twenty-two miles and running southeast by northwest. Fifteen degrees above this a light gray quartzite formation interlaced with slate and porphyry for a mile in thickness; then a vein of quartz that has been well established as fully twenty feet thick but is void of all minerals as far as yet known; above this another quartzite stratum, which is the foot wall of a silver –bearing galena ore, the largest I ever saw, being fully 10 feet of solid ore were exposed.

On Canyon creek, at the Oreornogo claim, this vein is now opened and being worked from the head of Beaver to Nigger prairie, on the South Fork, a distance of twelve miles air line. All of these before-named formations hold their respective places to each other, leaving no doubt as to their identity.

All the discoveries on this ledge justified the assertion that the history of silver mining does not show a vein as large, or long or well defined, as the one will speak for itself, being $125 per ton, fair average, and carries 40%, lead, making for each ton of ore 800 pounds of bullion.

To work this ore a smelter will be built at the mouth of Nine Mile creek (where five openings of the vein can be reached from five to six miles distant) and on the Mullan road twenty-eight miles from the Mission. [author's emphasis]

No mining camp was ever developed in one year. Leadville, Colorado, had been worked for ten years for gold before silver was discovered, and had the population of about 860. Three years later it had 40,000, with four smelters, four banks, and opera house that was the best in the state, schools and all that went to make a city prosperous. Its products made Denver a rich and beautiful city. The same can be said of Helena Montana, and the Virginia City, Nevada.

Silver ore means freight for steamboats and railroads, for freighters, work for coal burners and miners, smelters, and roads for an undeveloped section of Idaho.

Chapter IV

Founding and Losing the Wallace (Placer Center) Townsite

A detail of the 1887 photo of Wallace shown in its entirety on a later page. The area where Wallace was established was often referred to as a cedar swamp. This photo depicts numerous medium size stumps. The trees undoubtedly had previously provided a thick canopy that shaded the area and kept the ground moist, the conditions in which cedar thrives. However, there are no indications, such as ruts in the roadways, standing water, or corduroy roads, to confirm it was formerly an actual swamp. An arrow at the upper right points to the William Ross Wallace residence. *(Courtesy Butch Jacobson)*

Following the Civil War, Colonel William Ross and countless other veterans became part of the stampedes to all the major mining booms in the West. As addressed in the previous chapter, he came to the Coeur d'Alenes during the gold rush, which began after the news became public in the fall of 1883. (For a comprehensive account of the gold rush, see *The Coeur d'Alenes Gold Rush and Its Lasting Legacy* by Tony and Suzanne Bamonte.) Returning from the goldfields around Prichard and

Eagle creeks, he traveled the Mullan Road, by then in a sad state of disrepair. It was blocked by burnt or fallen timber and most of the bridges had washed out. He followed the South Fork of the Coeur d'Alene River to where he soon platted the town of Placer Center (later changed to Wallace). The site was it was a low, boggy area, often referred to as "The Cedar Swamp" because of the dense growth of cedar that covered it. The photos don't show a boggy area, as it had been ditched and drained. Efforts had also been made to deepen the waterway channels to improve drainage.

While the search for gold on the North Side was still actively underway, many prospectors fanned out in search of the source of the gold and other minerals. Many were combing the South Fork region by the spring of 1884. The first major lead-ore discoveries – the Tiger and the Poorman (soon consolidated as the Tiger-Poorman Mine) – occurred the beginning of May. They were located up Canyon Creek, which enters the South Fork of the Coeur d'Alene River at Wallace. Days later, on May 10, Wallace and some others located the Oreornogo claim in the same area. In addition to Canyon Creek, Wallace also staked claims on Nine Mile and Placer creeks. There was an immediate explosion of mining claims begin staked from that point on.

The New Town of Placer Center

In the spring of 1884, Colonel Wallace and a few associates built the first cabin at the future townsite of Wallace, which became the headquarters for mining exploration in the immediate vicinity. They then laid out a plat for the town and fenced in the 80 acres of land. On May 1, 1884, Wallace filed the plat for Placer Center, which was mistakenly believed to be in Kootenai County. For approximately a year, there was some confusion over whether the areas of Eagle City, Murray, and Wallace were in Shoshone or Kootenai County. During that time, the infamous Wyatt Earp lived in Eagle City, where he served as a Kootenai County deputy sheriff. In that capacity, he and his brother were involved in a shooting incident in Eagle City in April 1884. The newspaper reported they joked about the shooters' poor marksmanship, as the Earps were untouched by the flying bullets.

After filing the plat, Wallace quickly set to work building a town and trying to attract businesses and residents. A few additional cabins were put up within the next year or so, but building did not flourish until a sawmill, built by E. D. Carter, was established in 1886. As part of his development of the area, Wallace opened up a road, built bridges and trails up Nine Mile and Canyon creeks, and started the road that leads up Nine Mile Canyon from Wallace towards Murray. He later was responsible for completing the road over the divide at Dobson Pass.

History of North Idaho (Western Historical Publishing Company, 1903) offered the following summation of the town's development during the years when William Wallace was pouring his lifeblood into its survival and development:

> *The earliest pioneer in Wallace, in a business sense, was Alexander D. McKinlay, who came here April 16, 1885, accompanied by Peter J. Holohan, a partner with whom he has been associated twenty-eight years, in Idaho and other states. Mr. McKinlay recalls the fact that he was obliged to cross the swampy area of Cedar street by leaping from log to log and stump to stump in order to pass over dry shod. In 1886 Messrs. Howes & King located in "Placer Center" and opened a general store in a log building, having purchased the grocery business of A. D. McKinlay and J. P. Holohan.*
>
> *... The initial drug store vas opened by E. A. Sherwin, and the first hardware concern by J. R. Marks, William Hart, and E. H. Moffitt, whose pioneer institution is now the Coeur d'Alene Hardware Company. John Cameron arrived in the winter of 1886-7, and he became the proprietor of the original saloon on the townsite. The first business lot purchased from Col. Wallace was bought by E. D. Carter, in 1886, on which he erected a frame hotel.* [This was the Carter Hotel or House, of which M. D. Flint became the new proprietor after the hotel was completed in December 1886.] *The first livery and transfer business was controlled by Southerland & White.*

The Plat Map of Placer Center, Mistakenly Filed in Kootenai County

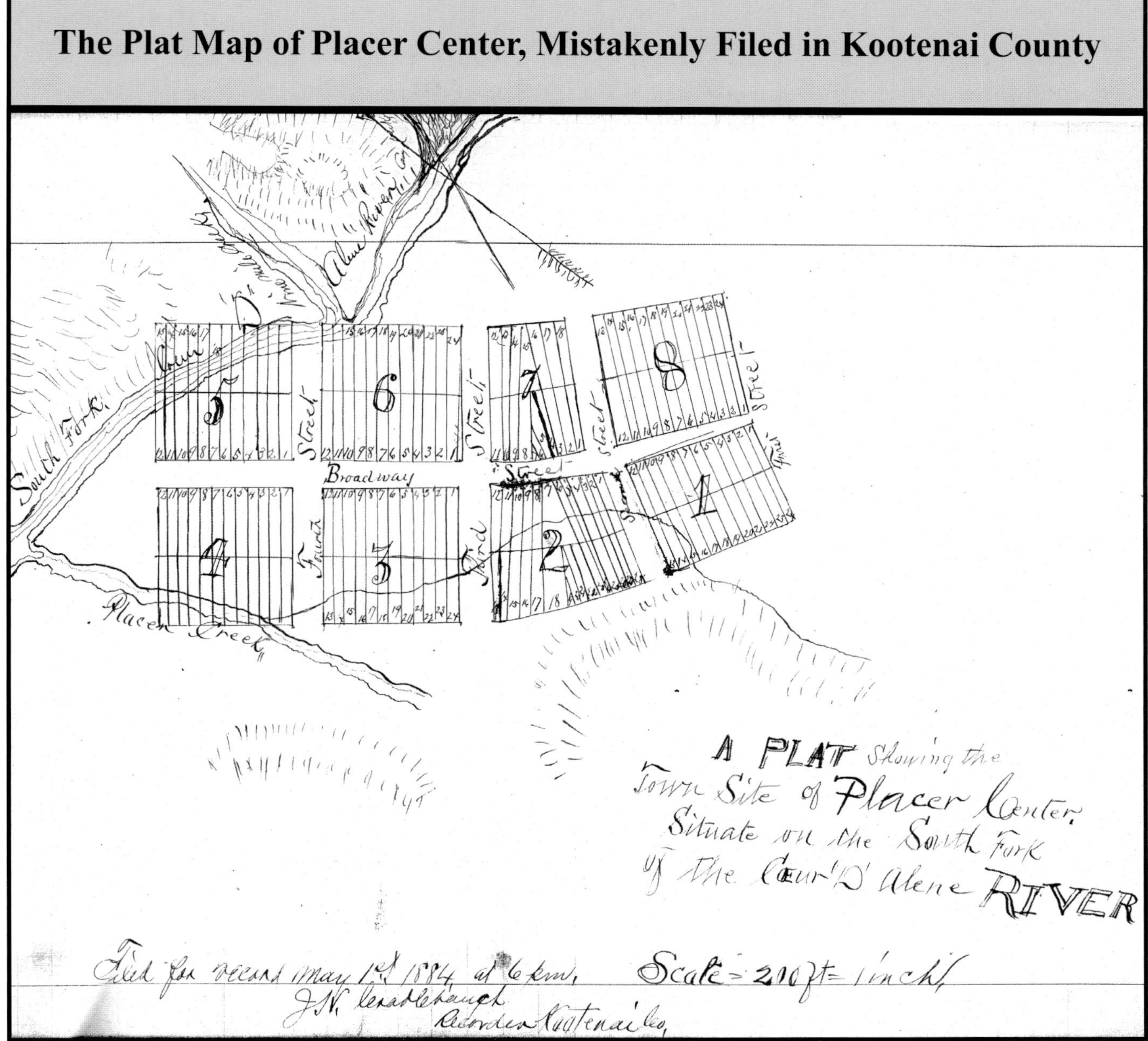

A copy of William Ross Wallace's original plat map of the "Townsite of Placer Center, situated on the South Fork of the Coeur d'Alene River," which was filed in Kootenai County at 6:00 p.m. on May 1, 1884. The site of Placer Center, as shown on this map, was at the confluence of the South Fork of the Coeur d'Alene River, Nine Mile Creek, and Placer Creek. In 1886, the name Placer Center was changed to Wallace, after its founder, William Wallace. Although the location of this townsite was in Shoshone County, it was filed in Kootenai County due to some confusion over the boundary line between the two counties during the gold rush period. The confusion extended to areas of Eagle City, Wallace, and Kellogg, which were believed by some to be in Kootenai County. Twelve days after the plat for Placer Center was filed, Thomas Irwin filed a plat in Kootenai County for his townsite at the foot of Jackass Trail, in what is now the Sunnyside district of Kellogg, where he owned and operated a boarding house and saloon. The confusion was the result of the boundary line not being clearly marked on the Idaho Territory maps by any known landmarks, simply because none existed. The matter was settled on July 11, 1884, by Judge Norman Buck. *(Map courtesy Kootenai County Recorders Office)*

A southwestern view of the fledgling town of Wallace, in 1887. The two main east-west streets with the boardwalks are today's Hotel and Bank streets. Bank Street the farthest north. Sixth is the other main street, which runs north and south. Colonel Wallace's house is perched on the hillside (arrow) above the end of Sixth, third from the top at the center of this photograph. E. D. Carter bought the first business lot from Colonel Wallace and built a frame hotel, which is the large structure on Hotel Street (closest to the hillside). Barely visible to the rear of the hotel is a catwalk to the double-decker outhouse (considered a modern convenience, as patrons did not have to walk down to the first floor to access the outhouse).The long, narrow building behind and to the left was the Carter Hotel barn. Carter also built the first sawmill in Wallace; stacks of wood are barely visible in front and to the left of Carter's buildings. On the next street opposite the Carter Hotel, the largest building is the Heller House. The post office is to the right and the Vedder & Co. general merchandise store is to the left. The printing on the next building appears to be "J.R. Marks & Co." which was the forerunner of the Holley, Mason, Marks & Co. (later the Coeur d'Alene Hardware Company). *(Photo courtesy Butch Jacobson)*

In 1887 Williams S. Haskins arrived from Kingston, with a fair stock of general merchandise. Subsequently he disposed of his goods and business to O. C. Otterson. The Dunn Brothers, A. J. and J. L., were the pioneer editors and proprietors newspaper owners in Wallace, their first venture being the Wallace Free Press.

In 1886, after Lucie (alternately spelled Lucy) Wallace became the town's first postmaster, the name Placer Center was changed to Wallace, as the original name was considered too long by the U.S. Post Office Department. That same year, Colonel Wallace and business partner Richard Lockey, a Helena, Montana, capitalist and co-owner of the Oreornogo, purchased Sioux half-breed scrip from a bank in Spokane. An application was then submitted for a patent on the Wallace townsite, and the scrip was given in payment. The two men, along with D. C. Corbin, the builder of the first railroad to the city, then formed the Wallace Townsite Company, of which Wallace was president, and had the town re-platted. As the town soon began to show some stability and growth, on December 4, 1886, Murray's *Coeur d'Alene Sun* printed the following:

The town of Wallace is more the holding its own in the way of solid and rapid improvement. One of the finest, if not the largest blocks in the Coeur d'Alenes is now in the course of erection by Col. Wallace. It is seventy-five by eighty feet in size. The ground floor will contain three spacious apartments finished in the very best manner, and the second will be devoted to a large public hall. Carter's sawmill cannot supply the demand for lumber. Flint's Hotel is to be enlarged. George & Human, of Delta, have bought two choice lots and intend building on them. Charles Seelig has purchased a location with a view of building a brewery thereon. Three saloons are doing a flourishing business. The work of clearing Main street is being pushed with vigor. The use of giant powder is clearing the town of stumps.

In early 1887, just as things were looking up for Colonel Wallace, he was notified about a problem with the Sioux scrip. The following was taken from Richard Magnuson's book, *Coeur d'Alene Diary*:

In 1887, the Colonel was called to Coeur d'Alene by the U.S. Land Receiver and informed that the scrip he used to buy the land was no good. Wallace claimed he then paid the land officer $50 "for advice" and was told the government's letter informing the land office about the scrip would not become a part of the Land Office records. Wallace then went to Spokane Falls to buy other land scrip so he could cover his purchase, but he found it was too expensive. He claimed the land officer told him to sell the land and no one could injure him for it. The land officer said he would protect him as his attorney.... Wallace contended the entry or issuance

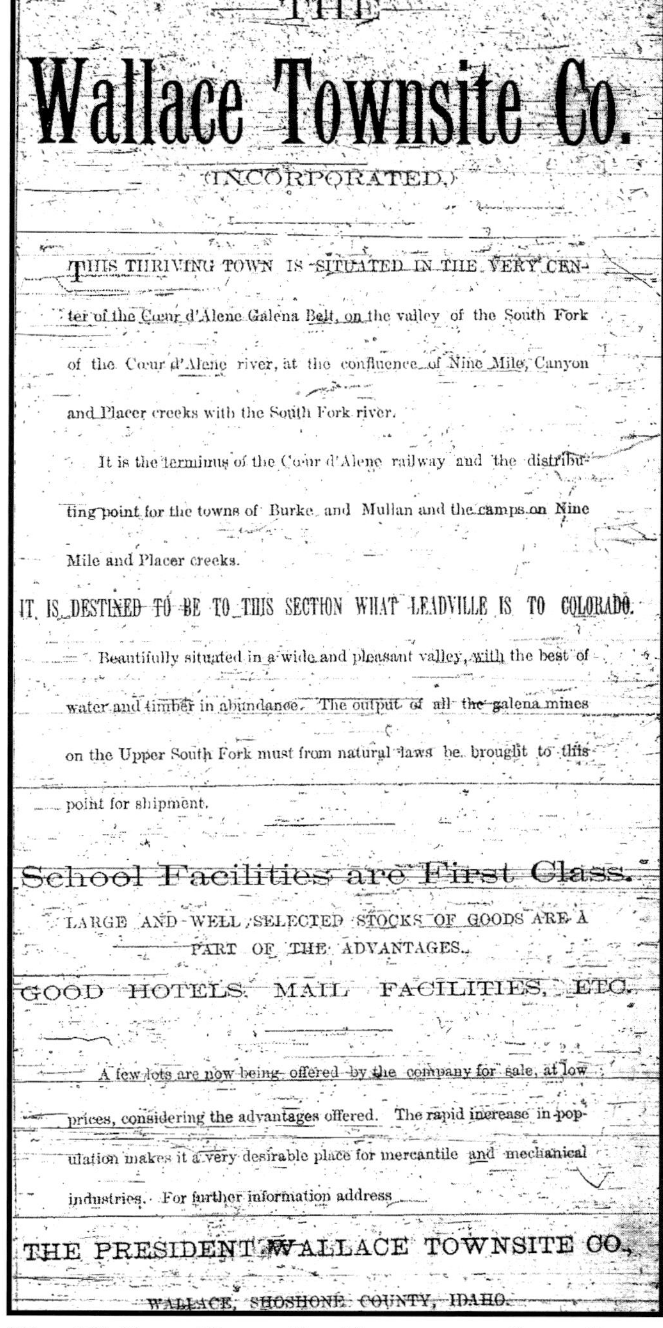

THE

Wallace Townsite Co.

(INCORPORATED.)

THIS THRIVING TOWN IS SITUATED IN THE VERY CENTER of the Cœur d'Alene Galena Belt, on the valley of the South Fork of the Cœur d'Alene river, at the confluence of Nine Mile, Canyon and Placer creeks with the South Fork river.

It is the terminus of the Cœur d'Alene railway and the distributing point for the towns of Burke and Mullan and the camps on Nine Mile and Placer creeks.

IT IS DESTINED TO BE TO THIS SECTION WHAT LEADVILLE IS TO COLORADO.

Beautifully situated in a wide and pleasant valley, with the best of water and timber in abundance. The output of all the galena mines on the Upper South Fork must from natural laws be brought to this point for shipment.

School Facilities are First Class.

LARGE AND WELL SELECTED STOCKS OF GOODS ARE A PART OF THE ADVANTAGES.

GOOD HOTELS, MAIL FACILITIES, ETC.

A few lots are now being offered by the company for sale, at low prices, considering the advantages offered. The rapid increase in population makes it a very desirable place for mercantile and mechanical industries. For further information address

THE PRESIDENT WALLACE TOWNSITE CO.,

WALLACE, SHOSHONE COUNTY, IDAHO.

The Wallace Townsite Company ad as it appeared in the *Wallace Free Press* on September 10, 1887.

of the duplicate scrip was fraudulent, and that he would fight to establish his rights.

On March 7, the town council met to consider ways to raise money to get a patent on the town land. Colonel Wallace asked that nothing be done for 30 days, as a land officer was on the way to investigate his land problem. His request was not complied with.

Wallace apparently believed the land officer and had faith that the issue could be rectified. According to the previous quoted passage, he did not try to conceal this from the city council, and the Wallace Townsite Company continued selling lots. Unfortunately, not only did the scrip issue continue to haunt him, eventually resulting in the loss of the townsite, but Lockey filed a lawsuit against him for the Oreornogo claims near Burke. According to local historian John Amonson, land at the northeast corner of the town of Wallace, near the eventual site of the Providence Hospital, was secured as a potential mill site, which could have played a part in Lockey's lawsuit.

In the fall of 1887, the first school was opened in a log building on the corner of Cedar and Third. In September, the first railroad into the mining district, the Coeur d'Alene Railway and Navigation (CR&N), reached Wallace. The arrival of the railroad was the single most important factor in the future success of the district's mines. Although the railroad's arrival was the pivotal turning point, the narrow gauge CR&N, built by D.C. Corbin and his associates, rapidly became obsolete and was swallowed up by the Union Pacific Railroad. Larger companies began competing for a piece of the action and soon built standard gauge lines. Wallace eventually became a railroad hub with the Northern Pacific servicing the area from the east to Wallace and the Oregon Railway and Navigation Co., predecessor of the Union Pacific, covering the area from the west to Wallace.

By November 1887, within two months of the railroad's arrival, Wallace's populated reached 500. It clearly was time to set up a city government. The following account regarding incorporation of the town is from *History of North Idaho*:

An 1889 photo of Wallace looking from the southeast to the northwest. This photo was taken around the time Colonel Wallace lost the townsite. *(Courtesy Butch Jacobson)*

At a meeting of the Shoshone county commissioners, May 2, 1888, J. C. Harkness presented a petition from citizens of Wallace praying for incorporation of the town, consisting of eighty acres, originally held by Colonel W. R. Wallace, by virtue of Sioux half-breed scrip. This petition was granted. The commissioners named as trustees: W. R. Wallace [the chairman, who essentially fulfilled the duties of a mayor], *D. C. McKissick, Horace King, C. Al. Hall and C. W. Vedder.*

The city of Wallace is located on portions of two sections, 34 and 27. The former section comprised a part of the original filing for townsite purposes, of Col. W. R. Wallace: the latter, railroad land owned, by the Northern Pacific railway company.

Wallace Townsite Jumped in 1889

Despite having purchased the Sioux scrip in good faith, then continued to act in good faith while diligently working to develop the townsite, the underlying problem did not get resolved. On February 19, 1889, by government order, the townsite of Wallace reverted from patented ground to public domain. An immediate land-jumping melee resulted, originally reported in the *Wallace Miner* and subsequently published as follows in *History of North Idaho*:

... hundreds who had paid for government title to lots found themselves with nothing but squatters' rights .... The excitement in Wallace that night was greater than on any other former occasion.

Lot jumping had grown to be quite an industry in our neighboring town of Mullan; it had thrived for a time in Burke, but no man had dared to squat upon a foot of Wallace. Sioux half-breed scrip had been placed on this eighty-acre tract for the sole benefit of the half-breed, Walter Bourke and his heirs, and the government would protect him in his right and title. W. R. Wallace and his company owned the land, and they had established an undisputed ownership. But when the Secretary of the Interior decided against a certain Sioux scrip location, near Glendive, Montana, a few of our citizens concluded that if that location was defective this one here must be.

It was not, necessarily, a logical conclusion, but it was sufficient for the purpose. One of the more daring looked over the town plat, and finally concluded that the lots on the corner of Cedar and Sixth streets possessed superior advantages as a business location, so he quietly walked over and took possession of it, posting a notice in a conspicuous place, asserting his claims. This was between nine and ten o'clock in the evening. This was the beginning.

The corner of Bank and Sixth was next taken. The jumping became general. Business men, laboring men, hoboes and rounders all joined the wild scramble for lots. Choice business locations went first, then outside property. Excitement was intense.

Hurriedly written notices claiming so many feet of ground were placed on every available lot. It was astonishing to see how excited some men got over the affair. A few were cool-headed and quiet in their demeanor. Notable among these was Henry Howes, who, viewing the wild uproar all around him, quietly said he had been working here hard for three years, and had finally secured a place to build a home. D. C. McKissick and C. B. Halstead were two other cool ones. They quietly took possession of three fine lots, at the corner of Cedar and Fifth, built a bonfire and sat up all night.

When morning dawned and our citizens realized the great change that had taken place in the ownership of property, it was regarded by some as a huge joke, so great was the confidence in the word of W. R. Wallace.

The hobo element was going to hold the ground, right or wrong. Possession was all they wanted, and this they had. The conservative element, which included the mass of the inhabitants, realized that the state of affairs was a serious thing for the town, so long as the status of legal right to the ground was unsettled.

What is Sioux Half-breed Land Scrip?

Land scrip and land warrants were certificates from the Government Land Office granting people private ownership of certain portions of public lands. Congress authorized issues of scrip – some directly, others only after trial of claims before special commissions or the courts. **Scrip was used primarily to reward veterans,** to give land allotments to children of intermarried Native Americans, to make possible exchanges of private land for public, to indemnify people who lost valid land claims through General Land Office errors, and to subsidize agricultural colleges.

The greatest volume of scrip or warrants was given to soldiers of the American Revolution, the War of 1812, the Mexican-American War, and, in 1855, to veterans of all wars who had not previously received a land bounty or who had received less than 160 acres. The next major scrip measure was the Soldiers' and Sailors' Additional Homestead Act of 1872, which **allowed veterans of the Civil War to count their military service toward the five years required to gain title to a free homestead. It also authorized those homesteading on less than 160 acres to bring their total holdings to 160 acres.** The government-issued scrip was in great demand as it could be used to enter the $2.50-an-acre reserved land within the railroad land grant areas and to acquire valuable timberland not otherwise open to purchase.

Other measures were enacted to indemnify holders of public-land claims that were confirmed long after the land had been patented to settlers. Claimants were provided with scrip equivalent to the loss they sustained. Indemnity scrip for some 1,265,000 acres was issued, most of which was subject to entry only on surveyed land open to purchase at $1.25 an acre. The chief exceptions were the famous Valentine scrip for 13,316 acres and the Porterfield scrip for 6,133 acres, which could be used to enter unoccupied, unappropriated, non mineral land, whether surveyed or not. These rare and valuable forms of scrip could be used to acquire town and bridge sites, islands, tracts adjacent to booming cities such as Las Vegas, or water holes controlling the use of large acreages of rangelands. Their value reached $75 to $100 an acre in 1888.

Least defensible of all the scrip measures were the carelessly drawn Forest Management Act of 1897 and the Mount Rainier Act of 1899, which allowed land owners within the national forests and Mount Rainier National Park to exchange their lands for public lands elsewhere. Under these provisions it was possible for railroads to cut the timber on their national holdings, then surrender the cut over lands for "lieu scrip" that allowed them to enter the best forest lands in the public domain. It was charged that some national forests, and possibly Mount Rainier National Park, were created to enable inside owners to rid themselves of their less desirable lands inside for high stumpage areas outside. The Weyerhaeuser Company acquired some of its richest stands of timber with Mount Rainier scrip. After much criticism, the exchange feature was ended in 1905.

By 1966, administrative relaxation had wiped out some distinctions between types of scrip; Valentine, Porterfield, and "Sioux Half-Breed" scrip were all accepted for land with an appraised value of $1,386 an acre, and Soldiers' and Sailors' Additional Homestead and Forest Management lieu scrip could be exchanged for land with a value from $275 to $385 an acre. At that time, 3,655 acres of the most valuable scrip and 7,259 acres of that with more limitations on use were outstanding.

The *Wallace Free Press* had a different perspective on the one-night event. They sympathized with the business owners who were apparently blindsided by the unexpected revelation. Suddenly, the land they thought was on patented land, and to which they held legitimate title, was now in jeopardy. As the paper pointed out, many of the so-called "jumpers" were the owners of those very businesses being threatened. No doubt, everyone agreed on the seriousness of the matter. Consequently, mass meetings followed and records were examined. It was – and remains – a complicated matter.

William R. Wallace's Rebuttal

On March 1, 1889, Colonel Wallace wrote the following letter to explain his position. It was printed in various local newspapers, including the *Coeur d'Alene Sun* and the *Wallace Free Press*. *History of North Idaho* printed a portion of the letter but, likely due to the sheer length of it, omitted some of the key elements.

For two years I held the land on which the town is built, on an agricultural location, and with my own hands split the rails, and segregated the same from the public domain; and when I had secured the title to the same by the location of Sioux half-breed scrip the outside fences were left standing and those on each side of the old Mullan road were taken down, as the title was unquestioned. And from all past precedents, as good as any patent could make it, the Department never claimed that the scrip was other than genuine, but in 1887, it made, through the General Land Office, a decision that because the scrip had been located by a duplicate in Dakota the original and only genuine scrip was cancelled. I can prove by the American Consul at Winnipeg, B. N. A., that the scrip was located by the original owner, Walter Bourke and wife; that he made oath, and still lives to verify the same; that he had never parted with the original and never gave any one power to use his name in any other location; never knew of any entry of this (his) scrip; and ***under this I felt perfectly secure that the title would be made to the Wallace Townsite instead of the fraudulent claimants in Dakota*** [author's emphasis], *and steps were taken to this end last November by D. C. Corbin* [one-half owner of the Wallace Townsite Company], *who employed the Hon. Luther Harrison of Washington, D.C., whose experience for over 15 years in the Department of the Interior, as associate counsel, makes his opinion second to none in this country, and in his report he says under date of Nov. 28, 1888:*

"The rule has been well established by Department practice, which has the sanction of the Supreme Court of the United States, that Sioux half-breed scrip may be located by agent duly authorized by power of attorney from the Indian, and that in case of unsurveyed lands the agent so authorized may make the improvements required by law."

"But I cannot understand by what line of reasoning an Indian is bound by a location made under fraudulent power of attorney; yet it will be observed that such is the line of reasoning of the decision canceling the location of Wallace." [author's emphasis]

"While I am not prepared to say that in any case his propositions are founded upon law, I do say that in this case he did not have the right to approve the location that was not authorized by law, or by the Indian, but was based upon fraudulent or forged powers of attorney. This being true, it appears to me that restitution for the loss sustained by the cancellation of the Wallace location and the detention of his scrip could be secured by bringing an action in the proper courts."

Now, I will not write you the particulars that I have in my possession concerning the local Land Office at Coeur d'Alene City, for I have furnished them, with documentary evidence, to the Associated Press of both the Eastern and Western departments, and it will become a matter for redress with Mr. McFarland if I am not correct. But for those whose local friendship I wish to retain I will say, that the jumping of the townsite and the entire interest, whether owned

This photo was taken in the fall of 1889. In regards to my earlier observation that there were no indications that this was actually a swamp – such as ruts in the roadways, standing water, or corduroy roads – a phrase "that it used to be a swamp" was still being used in 1889. According to John Amonson, who knows the history of the area well, he feels there is some truth to it being a swamp. He stated he was aware of a photo of the blasting of the river channel, which he assumed was done for better drainage. *(Courtesy Butch Jacobson)*

by those not here to defend their rights or not, has been general, and the town has gone into the hands (for the present) of men who have little regard for the personal or property rights of their neighbors. With this element the Wallace Free Press has found its level, and becomes the exponent of its kind. In connection with this I will say that when the proprietors of the Press first came here to look over the field for a newspaper, I made this contract with them: That to guarantee them success I would, and did, enter into an agreement with them, that their receipts should be $1,800 for the first six months, and to effect this E. D. Carter, myself and one or two others sent their paper broadcast over the United States and paid them the full price of their subscription, at the same time advertising with them in order to start the enterprise on a basis of pay. I also furnished them their office for six months free of charge, and at the end of that time sold them a lot which never belonged to the Town Company and the buildings thereon for what the buildings cost. They have been and are loudest in their expressions of any because their title is not good as they say it is not.

In response to the above portion of Wallace's letter, the Dunn brothers, owners of the *Wallace Free Press,* printed an article on March 16, 1889, in which they pointed out that both sides fulfilled their original agreement. However, Wallace was not alone in feeling the reporting of the townsite incident by the Dunn brothers was biased against him. The fact that John L. Dunn was among those to whom a townsite patent was granted in 1892 does raise some suspicions about having an agenda and the tool (the newspaper) with which to accomplish the desired result.

Following is the conclusion of Wallace's letter:

I have a partial list of those whose names appear on the lands of the townsite and those who have bought lands here in good faith, trusting in their title, and in due time I will publish them. I paid last year over $1,800 to regain the mine that I was first to discover on Canyon creek from the same class of people as those who have jumped this town, and on which I had previously expended over $4,000 in improvements. It is not for the credit of our Territory, or the community in which we live, that this is true.

I will close by saying that I have never received one dollar in the sale of town property which I have not expended in improvements thereon. I have built the streets, the bridges and furnished the money for the lumber that is in the sidewalks from the sale of town property, and the residents have received the benefits of the same. Property that two weeks ago would bring from $300 to $500 would not sell today for $10. I have paid this county over $2,000 in taxes on this townsite, and it has helped to build up our general interest.

I have built roads leading here, and borrowed money to pay for the same that this community might thrive. I have lived here through dark and gloomy days, when none would take part in the present townsite as a gift, and have been called a fool and crank because I could see in the future an opportunity to build a prosperous town, while they could not, but after having proven the prophecy of five years ago by my energy and work, they would rob me of what little I have left of the hardships and privations of pioneering.

The mob has tried my case without my being allowed a defense. The higher courts will ere long decide the validity of the claimants.

Respectfully yours. W. R. WALLACE.

Extended litigation followed. The owners of the Townsite Company continued to pursue a solution, still convinced the conclusion regarding the Sioux scrip was unfair and could still be reversed.

One Last Attempt to Retain His Townsite

In early 1891, Wallace filed a writ of certiorari (writ of review). For some unexplained reason, the request was denied, which just raises more questions about whether or not justice was ever served in this case. After the writ was denied on January 31, 1891, the February 7, 1891, *Wallace Press* reported:

The Town Trustees Take Steps to Secure a Patent for the Townsite. Col. Wallace and His Associates Were Denied a Writ of Review by the Secretary of the Interior

The Board of Town Trustees met on Tuesday evening at D. C. McKissick's store. All members were present except E. A. Sherman, who was reported sick. C. M. Hall was chosen chairman pro tem. Two dispatches for W. B. Matthews, the well known Washington [D.C.] *attorney, were read. One of these conveyed the information that the* ***Secretary of the Interior had declined to grant the writ of review prayed for by Colonel Wallace on the ground that the Land Office had erred in its decisions in the Wallace Townsite controversy.***

The other was a request from Mr. Matthews that immediate steps be taken to secure a patent. The first dispatch was made public last Saturday evening by a small "extra," and on Monday evening the Board of Trustees made arrangements with Frank C. Loring and James M. Porter, two deputy United States mineral surveyors, for a careful survey of the townsite, as a preliminary step for the application of a patent. The survey was to have been completed on Thursday. City Attorney Jones is now engaged in making out the remaining papers and collecting the necessary proofs. Mr. Matthews promises to procure a patent in sixty days. No other business was transacted by the Board.

The question of a patent for the townsite has been more or less discussed, but to within a few days ago it was not clearly understood. The decision of the Secretary of the Interior, just made known, has shut Col. W. R. Wallace and his associates out from any further claim to the townsite of Wallace.

Their last hope was the certiorari and this was denied them in an almost peremptory [high-handed] ***manner, Therefore the people of Wallace are at liberty to apply for a patent. Col. Wallace and his associates may adverse the application in the local Land Office, but it is the opinion of Attorney Jones and Mr. Matthews that the adverse would soon be set aside when the papers reached the Department at Washington, as the Commissioner of the Land Office could not reopen the case or ignore his own decisions.***

Hence it would be simply time and money wasted on the part of Col. Wallace to attempt an adverse. It is also said that E. D. Carter is in favor of patent, as neither his rights nor the claimants and possessors of what is styled "The Carter Block" will be affected one way or another. The case between Carter and the locators of what is known as his block is entirely different from that which existed between the people of Wallace and Col. Wallace, now settled for all time.

Now none but possessory titles prevail, but if a patent is secured warranty titles will be held. The cost of securing this patent will perhaps not exceed $3,000, two thousand of which go to the Washington attorney. The price of the land will only be $2.50 per acre, but there are so many incidentals that it will take economy to confine other expenses to $1,000.

With the explanations the attorneys give that there is no chance for further litigation or opportunity to reopen the case between Wallace and the people, the Trustees ought to proceed with due diligence to secure a title to title townsite from the Government.

Wallace Loses His Beloved Townsite

Although Wallace had secured the land through what was supposed to have been legitimate scrip issued by the government in exchange for Sioux lands, an unfortunate turn of suspicious events rendered Colonel Wallace's scrip worthless. After years of court proceedings, his claim to the townsite was successfully contested and, as a result, he lost not only the land but also his entire investment in terms of labor and associated development costs. In 1892, Colonel Wallace's townsite patent officially was denied. The ruling by the commissioner of the General Land Office was published in *History of North Idaho*:

By act of congress approved July 17, 1854, this class of scrip was authorized and the commissioner of Indian affairs issued the same. To the said Walter Bourke were issued five pieces of scrip for 480 acres of land. Said scrip, numbered 430, letters A. B. C. D and E., A and B for forty acres each, C for eighty and D and E for 160 acres each.

Upon representation made, that to 430 C had been lost, commissioner of Indian affairs issued a duplicate thereof; said duplicate was duly located March 9, 1880, and a patent for the land embraced in the location was properly issued as hereinbefore set forth. Thereafter, on the fifth of June, 1886, the register and receiver of the Coeur d'Alene land office, Idaho, allowed a location by W. R. Wallace, attorney in fact, with the original of said piece of scrip, for a tract of land, and on the seventeenth of said month transmitted the paper in the case for the action of this office in the matter. Said act of congress declares that no transfer or conveyance of any of said certificates, or "scrip, shall be valid."

In the case of Gilbert vs. Sharpson (14 Munn, 544) the court held that a "power of attorney as far as intended to operate as a transfer would be of no avail; the right of the half-breed in the scrip and the land would remain the same; it could not be made revocable nor create any intent in the attorney. Therefore the matter is solely between the government and scripee. It is claimed by the scripee that he did not locate the duplicate of said scrip and receive a patent for the land embraced in the location as herein stated, and the papers therein appear regular; therefore, it is reasonable to conclude that the location made with the duplicate scrip was properly made and that the patent therein was in full satisfaction of the claim of the scripee against the government as represented by the piece of scrip.

The government having thus discharged its obligation to the scripee, the original scrip was thereby rendered of no effect, and the location made therewith was fraudulent and void in its inception; and the cancelling of the same by this office as herein set forth, is authorized by the decision of the Supreme Court of the United States in the case of Harkness and wife vs. Underhill (Black, 316). Under date of March 13, 1889, the Register reports that Mr. Wallace was duly notified of the said decision of this office in this matter, but no appeal was properly taken therefrom. In view of the foregoing, and after a careful consideration of the arguments submitted by the counsel on both sides, I conclude that the action of this office was properly taken and the petition of Mr. Wallace is denied. Action is suspended under Rule 80, of practice.

A subsequent statement by a couple of bankers indicated that Sioux scrip was sold with the under-

John Amonson, Former Director of the Wallace Mining Museum, Addressed the Sioux Scrip Issue as Follows:

I struggle with the Sioux scrip issue. The more I think about it, the more I feel that things just don't add up.

What would prevent a half-breed with valid scrip from selling it "under the table," then disappearing to some other state or territory, and claim it was lost and seek to have duplicate scrip issued some years later? The half-breed would have received twice the value of the original scrip, and would have the Office of Land Management and/or the courts back him up with little or no investigation into the circumstances, that triggered the discrepancy. The net result would be outright fraud with the government being what amounts to be a willing participant. There is no clear evidence that fraud was implemented here, but the opportunity appears to exist.

There appears to be no redress of the situation William Wallace was in. There appears to be no fraud on his part, but he is denied any compensation whatsoever from using a government issued instrument of value. The evidence is overwhelming that Wallace ended up in this mess through no fault of his own, but experienced financial ruin as a result of alleged government incompetence.

standing that the banks would sell it at a fixed price, and the scrip was never guaranteed. Because it was disclosed that purchases were at the owner's risk, the banks had no subsequent liability. It appears, according to the March 9, 1889, *Wallace Free Press,* that Wallace had taken steps in investigate the validity of the scrip he purchased:

> *Colonel Wallace agreed to the* [bank's fixed] *price and ordered the scrip sent to Helena for inspection, stating that parties there would take an interest in it if it was all right. The Helena parties, after a considerable time, sent the money for it, which closed the transaction.*

The intervening years between the jumping incident and the final ruling denying William Wallace any rights to the townsite were tumultuous ones for both the townspeople of Wallace and its founder. Lucie Wallace filed for divorce, which became final on July 26, 1890, the day before the town went up in flames. All but a few of the wood buildings were destroyed, and the town became a tent city while it was being rebuilt, this time using brick construction.

William Wallace obviously was not one to be defeated by setbacks, such as he experienced with the Wallace townsite. As nearly as could be determined, he had moved to Spokane sometime in 1889, only returning to Wallace to take care of business. As mentioned earlier, for at least two years, he and former Civil War General A. P. Curry, along with Oliver Durant and M. E. Lindsey, operated a mining brokerage business, located in the Tull Block, under the name Wallace, Curry & Company. Later, on January 24, 1900, the *Weekly Journal-Miner*, an Arizona newspaper, reported on his present business engagement with the Yegor Canyon Copper Company:

> *Col. W. R. Wallace is superintendent of the property and he is doing the work thoroughly in a systematic manner and is building up a permanent camp. To accommodate the present necessities of the camp he has erected six buildings, all good substantial ones. The shafts are both good for their entire depth, and as far as development work the indications are good for making a first-class mine. Development work will be continued to a depth of at least 600 feet . . .*

Apparently, however, Wallace was in failing health, and for that reason, he and Annie moved to Whittier, California. It was only one year later, on November 16, that he died. He was 67 years old.

Despite studying the numerous newspaper articles regarding the townsite patent, a clear picture of who exactly was at fault remains elusive. Errors seemed to have occurred all the way up and down the chain. This begs the question of why others, as far as I could tell, were not held accountable. Until receiving the final ruling, it appears that Wallace and his business partners continued to hold to the belief the townsite issue could be legally resolved and, acting in good faith, worked toward that end. As it turned out, Wallace lost everything, including his wife, Lucie (see "Wife number two," starting on page 31).

The Wallace Townsite Patent

The townsite of Wallace finally received a patent on June 11, 1892. **The following detail of who received the patent** is from *History of North Idaho*:

"On June 11, 1892, a townsite patent for the **Wallace location was issued to John L. Dunn [trustees chairman and newspaper owner],** John B. Cameron, George P. White, Henry E. Howes and Thomas A. Helm."

It is common knowledge that the press has always held tremendous power over politicians. Depending on the spin they apply to their news stories the press can make or break a politician. In 1846, the Associated Press was founded. Consequently, once a story has been written, that same story is available throughout the nation. It would have been in any politicians best interest to cater to the press, even if it meant causing harm to an innocent and undeserving person. It would also be in the best interest of the press to cater to a politician if that politician were able to help someone who owns a newspaper and wants a return favor. I found it suspicious that the owners of the newspaper ended up with part of the Wallace townsite.

Colonel Wallace's last business venture in the Coeur d'Alenes
Purchase of a steamboat involved in the worst steamboat tragedy

Although steamboats plied the waters of Coeur d'Alene Lake and its primary tributaries from 1880 until the 1950s, the most tragic boating accident during that era happened just below the landing at Kingston. Monday, April 4, 1887, was the fateful day when the small steamer *Spokane* capsized during a trial run in preparation for putting it on a regular schedule from the town of Coeur d'Alene to the Old Mission Landing. The unfortunate circumstances under which the accident happened included an uncertainty about how many passengers were aboard. Five bodies were recovered, but there was concern that at least one more person aboard the steamer had drowned, though no other bodies were ever found. The five deceased men were known around the area, but the most shocking death was that of J. C. Hanna, the highly respected Spokane Falls city clerk.

The steamboat had been purchased in November 1886 by Nelson N. Martin, who had already gained favor in the area as the owner of the Spokane Falls and Coeur d'Alene Stage Line. Shortly after moving to Spokane Falls from Truckee, California, in early 1885, he established the first regular stagecoach service to the North Side district. Like Corbin's railroad, Martin's stagecoach line made a steamboat connection from the town of Coeur d'Alene to the Old Mission Landing. Martin initially contracted with the owner of the steamer *Coeur d'Alene*. Procuring his own steamboat, which he named *Spokane*, was a way to expand his transportation business.

The *Spokane* had been purchased from George Ellis, who stayed on as the pilot and engineer. (One newspaper account indicated Ellis either had been or presently was a co-owner of the boat, but other accounts named Martin as the sole owner.) The thirty-foot boat was built on the Snake River in 1882 and had only recently been moved to Coeur d'Alene Lake. It was described as being "the best boat of its kind in the Northwest."

According to the first report of the accident, printed in the April 6, 1887, *Spokane Falls Morning Review,* "the boat was running at a high rate of speed when she struck a drift" on the island a short distance below Kingston and "swinging off, capsized." It went on to say, "The passengers were thrown into the river and most of them managed to escape, and although every effort was made to save those in the water by the men first reaching shore, five or six sunk from sight." It was estimated that nineteen to twenty-one passengers were aboard, but no freight. Concerning questions about the boat and an implication it was overloaded, Martin responded:

> *. . . In conclusion I will say that no man knows what I have suffered or endured in the past few days . . . I had forgotten to say that I was arrested at Kingston by the U.S. Marshal, but I explained my position to him and he at once released me. The people of Spokane Falls and a large number of inhabitants of the Coeur d'Alene mines well know how I have served them in the past two years as a public carrier. I have always aimed to have careful drivers and to please my patrons and do the best I could by all, and am grieved to see so many false rumors about myself. With the above statement I feel assured that no one will attach any blame to me for the sad accident, and I thank God so many of us were saved. I also will say that at the coroner's inquest I was exonerated from any blame in connection with the accident.*

Following the accident, knowing the *Spokane* would be forever tainted as a passenger boat, Martin put it on the lake as a freighter. He and a partner bought another passenger boat. Unfortunately, the freighting operation was not successful and subsequently moved to the St. Joe River. **In 1889, the Spokane was sold to Col. William R. Wallace, founder of the town of Wallace, who renamed it the Irene.** Wallace operated the Irene on Lake Coeur d'Alene until he disposed it. Nelson Martin and his wife, Mattie, eventually moved to San Diego, where he died on February 3, 1930, at age 82. The most unfortunate part about this accident was that he had been known for his reputable service, and the accident occurred while in the performance of a kindly service for no personal gain.

The following article was taken from the *Idaho Statesman* of January 17, 1890. It was one of Wallace's final acts during his transition between Spokane and Wallace.

Coeur d'Alene Reservation
Coeur d'Alene Times

Friday, January 3rd, 1890, is the day set for the grand "pow wow" of Coeur d'Alene chiefs and representatives of Uncle Sam concerning the ceding to the government of a very large portion of the Coeur d'Alene Reserve, all of which is in Kootenai county.

Col. W. R. Wallace, whom Capt. John Mullan, attorney for the Indians, recently appointed deputy attorney, with full power to act, and Chief Saltese and sub-chiefs will represent the Indians, while his honor, Judge Willis Sweet, and others will be present on behalf of the government.

It is a matter of speculation as to whether or not the Indians will sign the papers (not in themselves considered binding being required as an expression of will) until $50,000, which the government promised to pay them in a treaty made in 1878, is forthcoming. They may naturally mistrust the government and refuse to make even promises until former agreements are fulfilled.

Should the conference prove successful, without some unforeseen delay, the ceded portion of the reservation would be opened to settlers some time next spring. Hundreds of families will rush into Kootenai county to secure valuable and worthy homes and hundreds of prospectors will here find room to ply their tireless and patient vocation for years to come.

How it played out

The Tribe knew they were to be paid for the reservation lands as then defined. However, the 1889 Indian Appropriations Act included a provision directing the Secretary of the Interior "to negotiate with the Coeur d'Alene tribe of Indians," and, specifically, to negotiate "for the purchase and release by said tribe of such portions of its reservation not agricultural and valuable chiefly for minerals and timber as such tribe shall consent to sell." Later that year, the Tribe and government negotiators reached a new agreement under which the Tribe would cede the northern portion of the reservation, including approximately two-thirds of Lake Coeur d'Alene, in exchange for $50,000.

The new boundary line, like the old one, ran across the lake, and General Simpson, a negotiator for the United States, reassured the Tribe that "you still have the St. Joseph River and the lower part of the lake." And, again, the agreement was not to be binding on either party until both it and the 1887 agreement were ratified by Congress.

On July 3, 1890, while the Senate bill was under consideration by the House Committee on Indian Affairs, Congress passed the Idaho Statehood Act, admitting Idaho into the Union "on an equal footing with the original States," The Statehood Act "accepted, ratified, and confirmed" the Idaho Constitution, which "forever disclaimed all right and title to all lands lying within [Idaho] owned or held by any Indians or Indian tribes" and provided that "until the title thereto shall have been extinguished by the United States, the same shall be subject to the disposition of the United States, and said Indian lands shall remain under the absolute jurisdiction and control of the congress of the United States."

A little over a month later, on August 19, 1890, the House Committee on Indian Affairs reported that the Senate bill ratifying the 1887 and 1889 agreements was identical to the House bill that it had already recommended. On March 3, 1891, Congress "accepted, ratified, and confirmed" both the 1887 and 1889 agreements with the Tribe. The Act also directed the Secretary of the Interior to convey to Frederick Post a "portion of [the] reservation, that the Tribe had purported to sell to Post in 1871. That property, located on the Spokane River and known as Post Falls, was described as "all three of the river channels and islands, with enough land on the north and south shores for water-power and improvements."

Chapter V

The Beginnings of Wallace

A steam engine and two passenger cars of the Northern Pacific Railroad just outside of Wallace, circa 1888. *(Courtesy Butch Jacobson)*

The following is a comprehensive description of Wallace from the January 1, 1901, *Coeur d'Alene Mining Journal*, as it was printed in *History of North Idaho*:

Wallace's First Major Fire

The first serious loss by fire in Wallace occurred Sunday evening, July 27, 1890. This overwhelming disaster originated in the Central Hotel, on Sixth street. When the end came Wallace was, practically, in ashes. Ten minutes' service by the fire department resulted in exhausting the water supply, and the voting city was at the mercy of the flames. Fanned by a stiff gale they spread up Sixth to Cedar street, leaped Cedar; and in a few minutes later the Hanley House and Club theater were ablaze. From this period the fate of

Wallace Idaho in 1888. This first Wallace depot is in the foreground with the Coeur d'Alene Railroad and Navigation Company engine and car to the left. *(Thomas Elsom photo collection, courtesy Dean Ladd)*

Wallace before the great fire of July 27, 1890. *(Courtesy Butch Jacobson)*

the doomed town was assured. With the single exception of one building six blocks were destroyed, the one solitary edifice remaining in this section being the Pavilion, corner of Cedar and Fifth streets. Giant powder was brought into service to check advancing flames by blowing up buildings, but such efforts proved futile. So rapid was the work of destruction that absolutely nothing of the immense stocks of goods was saved. One fatality resulted: An Italian, Centinmio Denarco, was burned to death while in a drunken stupor in the New State saloon, on Sixth street. Within the boundaries of Fifth street the river on the east and north, and the hills to the south, every business house and residence was destroyed. The loss, as estimated by the Murray Sun, was $500,000, with insurance of only $43,750.

In speaking of the generous proffers of aid the Murray Sun, of July 30, 1890, stated: The towns of Mullan, Wardner and Osburn being on the line of railroad; and in easy communication with Wallace, sent car-loads of provisions early Monday morning. Offers of assistance were telegraphed from Spokane and other towns, but were declined with thanks, the surrounding towns being amply able to relieve the temporary necessities of the people. The disaster was an appalling one, but not enough to injure the town permanently, as the work of rebuilding will be on a larger scale than before. Petitions have been presented to the board of trustees of Wallace, praying that only iron, stone and brick buildings may be erected in certain down town districts, and these petitions have been granted. Thus far the Coeur d'Alene Clothing Company is the only one that has failed, as a result of the fire."

In June, 1892, the trustees purchased six lots on the northeast corner of Third and River streets, comprising an area of 100 x 150 feet, and a most eligible site for school house purposes. The price paid was $1,750. The same year a handsome brick building, two stories in height, surmounted by a tower of Moorish design, was erected. During the spring of 1901, an annex to the original edifice was built, somewhat larger, but of the same style of architecture and general design. It is, at present, one of the most attractive in the city. The excellent school facilities at present enjoyed by Wallace are developed from humble and primitive origin. In the fall of 1887 the first school was opened in a log building on the corner of Cedar and Third streets . . . Here school was continued for one year, when a frame building was exchanged for a log shack on Pine, between Fifth and Sixth streets. One year later the trustees of the school district erected a building for school purposes which was subsequently converted into a residence by O. B. Olson. From this institution, on May 17, 1895, was graduated the first class in the history of the Coeur d'Alenes. The exercises took place at the Methodist Episcopal church under the direction of Prof. C. W. Vance. The graduates were Myrta Howes, Nina Hogan, Katie Hanley, Katie Baldwin and Luneti Worstell. With the exception of two slight epidemics of diphtheria the Wallace schools, in District No. 8, enjoyed uninterrupted success during the term time since 1895. At the present writing, 1903, the officers and faculties of the grades are as follows: Superintendent, H. M. Cook; principal of high school, Mrs. Edna Clayton Orr; music teacher, Miss Grace Jenkins; grade teachers, Kathryn Cunningham, Mollie Fulmer, Myrta Howes, Sadie Skattaboe. The high school enrollment in 1903 was, boys, eighteen; girls, twenty-four. The average attendance is twenty-nine. During the past nine months there have been no graduation exercises, Charles Dunn, eligible to that honor, having been appointed to the naval academy, Annapolis, and dropping out without formal graduation from the high school.

A description of Wallace in 1900

A southeastern view of Wallace after a devastating fire on July 27, 1890 that destroyed virtually all of Wallace's business section and many homes. Fortunately, there was only one fatality. According to *History of North Idaho*, "An Italian, Centinmio Denarco, was burned to death while in a drunken stupor in the New State saloon, on Sixth street "The long building in the foreground is the CR&N depot situated along the South Fork of the Coeur d'Alene River. Compare this photo with the previous one to get a sense of how much growth Wallace had experienced in just three years. Almost the entire burned area had been crammed with buildings. The day after this fire, the towns of Mullan, Wardner and Osburn, nearby communities along the railroad line, sent carloads of provisions to assist the residents of Wallace. Spokane and other towns telegraphed offers of help, but since the surrounding towns had been so generous, these offers were declined. Within four days of the devastating fire, a tent city had sprung up and before the year was out many of the old buildings had been replaced by brick structures. *(Courtesy Butch Jacobson)*

The following comprehensive resume of existing conditions in Wallace in 1900 was furnished by the Coeur d'Alene Mining Journal, of date January 1, 1901 gave to Wallace its greatest measure of advancement for a single year in material improvement and general enlightenment, and her citizens cross the thresh-hold of a new year satisfied with the treatment they have received at the hands of the departed year. . . . Suffering nearly total destruction by fire on July 27, 1890, the town has made better progress than formerly, until it has become what the severe critic would term an "ideal mining town," possessing all the embellishments of modern civilization, splendid buildings, superb electric light and water plants, all the standard fraternal organizations and re-

City of Wallace, Idaho, City Chairmen and Mayors

Founded May 1, 1884
Incorporated on May 22, 1888
Prepared by Wallace city clerk, Joanne McCoy

Chairmen:

May 1888 - March 1889 – W. R. Wallace
April 1889 - April 1890 – C. W. Vedder
April 1890 - April 1891 – E. H. Sherwin
April 1891 - April 1893 – J. L. Dunn

Mayors:

April 1893 - April 1894 – W. S. Haskins
April 1894 - April 1895 – Oscar B .Wallace
April 1895 - April 1897 – Jacob Lockman
May 1897 - April 1898– Herman Rossi
May 1898 - April 1899 – T. N. Barnard
May 1899 - April 1902 – T. F. Smith
May 1902 - April 1904 – T. D. Connor
May 1904 - April 1907 – Herman Rossi
May 1907 - April 1909 – Hugh Toole
May 1909 - April 1911 – Walter H Hanson
May 1911- April 1915 – James Taylor
May 1915 - April 1917 – Charles Mowery
May 1917 - April 1919 – Homer Brown
May 1919 - April 1927 – Hugh Toole
May 1927 - April 1929 – W. H. Herrick
May 1929 - June 1930 – Herman Rossi
July 1930 - April 1931 – J. H. Munson
May 1931- May 1935 – Emil Pfister
June 1935 - April 1937 – Herman Rossi
May 1937 - April 1943 – Lawrence Worstell
May 1943 - April 1949 – John Batts
May 1949 - April 1951 – R. G. Binyon
May 1951 - April 1957 – C. A. Magnuson
May 1957 - July 1957 – George Albertini
July 1957 - April 1959 – Clyde Murray
May 1959 - December 1973 – Arnold Keller
January 1974 - March 1980 – Herbert Wellman
April 1980 - February 1981 – Paul Decelle
March 1981 - December 1985 – Frank Morbeck
January 1986 - Nov. 1991 – Maurice Pellissier
December 1991 - Sept. 1992 – Gregg Kimberling
October 1992 - March 1996 – Debbie Mikesell
April 1996 - December 1997 – Archie Hulsizer
January 1998 - December 2009 – Ronald Garitone
January 2010 - Current – Dick Vester

JoAnne McCoy, Wallace city clerk, in her office. She personally hand sorted the Wallace minutes to find the names and dates of all the mayors. *(Courtesy Jamie Baker)*

ligious societies, good public schools, a magnificent Thespian temple, etc. The city's population today is only two thousand, but the next five years assures an increase of not less than five thousand, and proportionate expansion along other lines.

Wallace made permanent progress during 1900 in every commendable respect, truly reflecting the unprecedented prosperity in local mining circles. The municipal improvements, new cross walks. Leveling, grading and macadamizing of streets, made during the year, cost $14,450.29, every dollar of which was judiciously expended. The city officials, led by Mayor Smith, have made a splendid record. Municipal affairs have been well managed, and let it be said incidentally that the city's indebtedness is only $8,000, a bond obligation created to furnish the city with a first-class sewer system. And it should also be stated in this connection that the Wallace Manufacturing, Electric & Power Company has a splendid combination light and water plant, which has a patronage of 3.000 incandescent and eighty arc lights, and 600 faucet and surface connections. The double plant is one of the best and most perfectly equipped in the west, worth $150,000, and is under the management of H. W. Fellows, the officers of the proprietary company being A. B. Campbell, president; Richard Wilson, vice-president and F. F. Johnson, secretary and treasurer.

The business of the Wallace post office for 1900 was one third greater than for the previous year, which very accurately reflects the general advancement of the community. According to comparative population of Idaho towns, the Wallace post office easily ranks first.

But the record of the First National Bank is another evidence of the permanent advancement of the city. By the official report of the First National, issued December 13, it was learned that the individual deposits were $552,313, against $466,523 for the corresponding month of the previous year, while the deposits, subject to check, in September, 1898, were only $272.703. The First National began business August 8, 1892, and closes the year ago with resources of $784,513.

The building record for Wallace in 1900

The building record made by Wallace during 1900 eclipsed that of any preceding year, the money expended in the erection of business houses reaching close to the $100,000 mark, while the erection of dwellings and handsome cottages absorbed about $50,000.

The new business structures included the Holohan-McKinlay block, two-story brick, with fine store rooms on the first floor, $20,000. Sunset brewery, four-story brick block, erected and owned by ex-Mayor Jacob Lockman and association, $15,000; brick warehouse, by White & Bender, $12,000.

Coeur d'Alene Hardware Company, warehouse No. 2, facsimile of No. 1, $7,000; Furst block, two-story brick, by John Furst. $7,000; Otterson, two-story brick, 50 x 100, by O. C. Otterson, $12,000; Heller, two-story brick block, by Mrs. Eliza Heller, $7,000; Carl Mallon brick and stone brewery, $6,500; Mayor P. F. Smith, warehouse, $3,500; Jones & Deane, addition to second story, $5,000. Coeur d'Alene iron Works, two-story- frame, 35 x 100, $3,000, fifty feet addition; two-story, to Odd Fellows Hall, $3,000; Wood & Keats, two-story frame, 25 x 60, $2,500; Al. C. Murphy, lodging house, $2,500; Fred Kelly, lodging house, $2,500.

George F. Moore, improvements on furniture store, $2,000. Total, $97,000.

The new residence buildings included: William Hart, $7;500; E. Protesting, $6,500.; Dan McGinnis, $5,000; Mrs. Moriarty, $3;000; George Garrett, $3,000; W. D. Powers, $2,000; and John Pressley, $8,200.

Among the business houses represented in the Wallace Press in November, 1890 were:

Hotels: Carter House, E. D. Carter, proprietor; the Idaho, N. R. Penny; the Crazy Horse, Simnett & Webster; Hanley House; American House, George H. Heller; Michigan House, Charles Mehl.

Liveries: Sutherland & White, Red Front, McDonald & Johnson. Clothing houses: the O. K.. Julius Kohn, manager; the Colorado, Sam Heller.

Restaurants: the Cedar Street, Mesdames Hogan & Place, proprietors; the Frankfort. City transfer ,Paul F. Smith, William H. Otto.

Jeweler: Eli Ritchott.

Meat markets: Silver Belt; Barger & Sears, proprietors; Follet & Harris.

Hardware: Holley, Mason Marks & Company. Furniture and undertaking: William Worstell.

Bank: Bank of Wallace, Charles Hussey, proprietor, C. M. Hall, cashier.

Ice: Carl Mallon.

Dry goods, Mrs. A. A. Scofield; Galland Brothers; Chicago Beehive.

Blacksmith: James Hennessy.

Real estate: N. Witner.

Bakeries: City Bakery; Muir & Dicks, proprietors; Wallace, Woods & Keats; Bank

Street Baker; Paul Herlinger & Company.

Tin shop: K. B. Sauter, proprietor.

Photography: T. N. Barnard's Studio.

Lawyers: Walter A. Jones and J. C. Harkness: Henry S. Gregory; W. B. Heyburn ; A. L. Dunn.

Building contractors: Fuller & Warren.

Cigars and tobacco: W. H. Leghorn & Brother.

Sash and door factory, H. Wood.

Painters: McFadden Langrell.

News stand. Tabor & Vinos.

Harness shop: J. M. Carmelius.

Justice of the peace: A. E. Angel.

Doctors: U. T. Campbell. A. Boston.

Engineers: W. Clayton Miller; George R. Trask.

Grocers: White & Bender; Howes & King.

Tailors: Pfister & Wassenberg.

Wholesale liquors: D. C. McKissick. Brewery, Carl Mallon.

Street lighting: Wallace Manufacturing; Electric & Water Company.

Newspapers: The Wallace Press; the Wallace Miner.

Bottling Works: Staley & Zweifel.

Harness and saddlery J. M. Carmelius.

There was also the American Loan Company.

The saloons of Wallace at that period numbered twenty-seven.

General merchandise was represented by Dobson & Nottingham.

The Wallace Board of Trade, in March, 1902, completed all necessary details for the establishment of a publicity bureau, taking cognizance of the mines of this particular district. To prospectors and small property holders blank forms were forwarded, with a request that as full in-

formation as possible be furnished the bureau concerning mines, prospects and other holdings. Responses have been full and complete, and the enterprise is a commendable success. Free postal delivery is not yet a fact in Wallace, but should the increase in post office business for the succeeding two quarters equal the last two, free delivery will, undoubtedly be established.

The census of 1900 gave Wallace a population of 2,265, and today it is claimed, on fairly substantial authority that there are within its limits 3,000 inhabitants.

Early in the morning of November 11, 1898, the Idaho Hotel, John B. Cameron's and Thomas Reynolds' saloon buildings, all two-story frame structures, were destroyed by fire, and the Fuller House badly damaged, involving a loss of $12,000, with insurance of only $500 on the Idaho Hotel's furniture. The latter property, owned by Glen McDonald, and leased by Johnson & Wilmot, contained forty-one rooms, all of which were occupied. In this disaster a rare quality of heroism was displayed by Gus Enz, the night clerk. Discovering the fire he ran up stairs and awakened the sleeping guests, all of whom, with the exception of John F. Moore, a waiter, and W. H. Dwyer, a cigar maker, escaped. These two, and Enz, the faithful, courageous night clerk, were burned to death, the latter laying down his life to save others. Mrs. Alice Finnegan, chamber maid, George W. Mitchell and William Palmer were seriously burned.

As in the case of the great fire of 1890, immediate steps were taken to repair the damage sustained in the disaster of 1898. In both events Idaho pluck and energy were in evidence, and scarcely were the ashes of the 1890 conflagration cold that temporary buildings sprung up to be soon replaced by more substantial structures.

The sentiment of the people of Wallace concerning the remarkable progress in rebuilding the town is fittingly voiced by the editor of the Free Press in the issue of November 29, 1890: "Yesterday, four months ago, there were few men in Wallace who looked forward to such a cheerful and happy Thanksgiving as the one enjoyed on Thanksgiving last. On the twenty-eighth of July the town presented a scene of desolation. On Thursday last there was a rebuilt city; with substantial brick blocks and frame buildings from

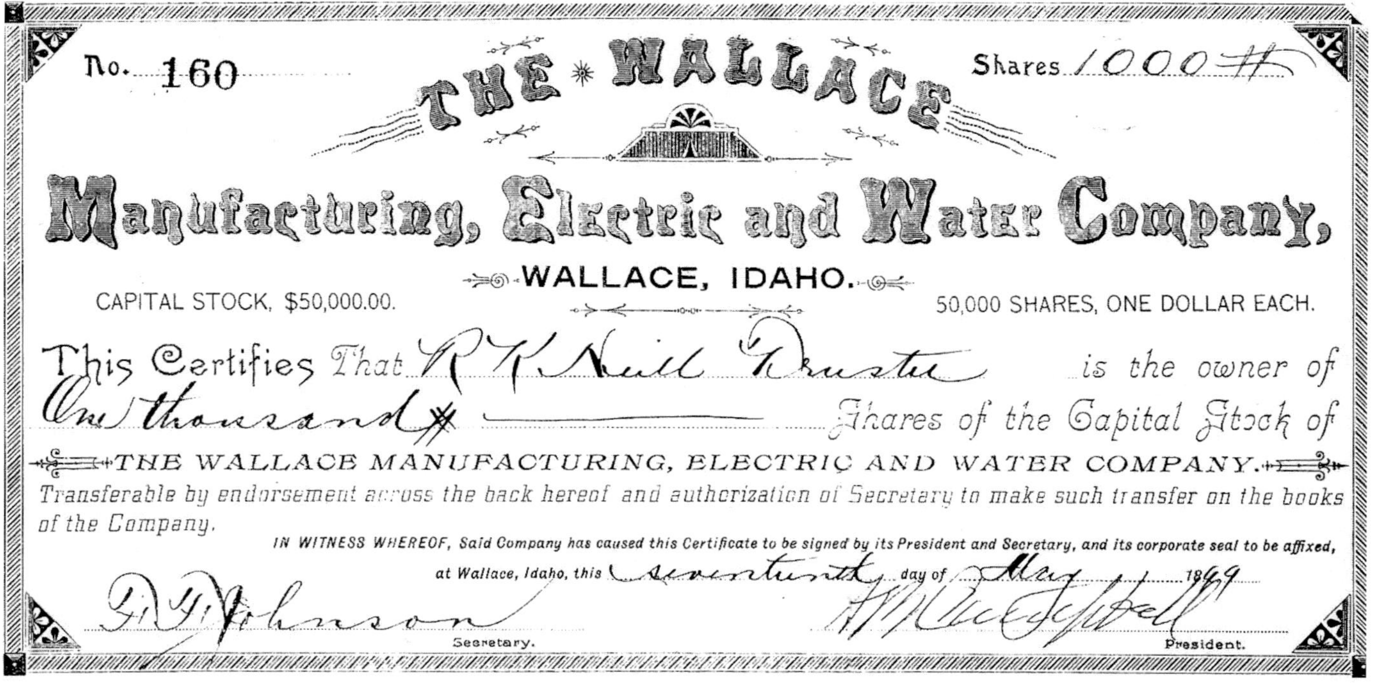
No. 160 Shares 1000 #

THE WALLACE

Manufacturing, Electric and Water Company,

WALLACE, IDAHO.

CAPITAL STOCK, $50,000.00. 50,000 SHARES, ONE DOLLAR EACH.

This Certifies That R. K. Neill Trustee is the owner of One thousand # Shares of the Capital Stock of

THE WALLACE MANUFACTURING, ELECTRIC AND WATER COMPANY.

Transferable by endorsement across the back hereof and authorization of Secretary to make such transfer on the books of the Company.

IN WITNESS WHEREOF, Said Company has caused this Certificate to be signed by its President and Secretary, and its corporate seal to be affixed, at Wallace, Idaho, this seventeenth day of May 1899

Secretary. President.

one end of the burnt area to the other, covering nine squares. In the face of limited transportation, lack of building materials and labor suitable for the work to be done, the recovery from a bed of ashes in so short a time borders on the magic. The transformation is complete.

Wallace is today a more substantial city than before the fire; more solidly built and more thrifty. Much of this is due to the handsome brick buildings that have been erected, which stand as monuments of enterprise in our pretty mountain home. Among the most prominent of these edifices are the Hardware Block, erected by Holley, Mason, Marks & Company, D. C. McKissick's wholesale liquor block, Colonel Steward Fuller's new hotel, the National Bank block, the Bank of Wallace block, L. Manheim's block, Howe & King's, O. C. Otterson's, White & Bender's and Mrs. A. A. Scofield's blocks. These are all new brick blocks, modern and completely furnished. Their cost is not less than $100,000, which is very good testimony to the vitality of the business men of Wallace and their unshaken faith in the grand future of the town.

Following the terrible trial by fire of July 27, 1890 a new hose company was organized to replace the old fire department, which had been practically innocuous since that event. It was named Wallace Hose Company No. 1, and comprised a membership of twenty active men, with Adam Aulbach as foreman. It came into being at a public meeting held November 2, 1890. Of the old department Scott McDonald had been chief. The new company comprised, aside from Foreman Aulbach, Scott McDonald, first assistant, A. P. Horton, second assistant, Julius Kline, Jacob Lockman, James Hennessy, John Frazer, Pipe men; Peter Holohan, Jesse Tabor, hydrant men; George Heller, H. D. Sawyer, Ed Sarbin, Louis Kosminsky, Charles Woodman, O. C. Otterson, Al Honeke, Harry Germond, A. H. Utley, Angus Sutherland and J. M. Carmelius, Hose men. The present organization is partly a volunteer department, of which Fred H. Kelly is chief and M. C. Murphy assistant. The excellent water system was established in 1890 water being led from Placer creek, the reservoir being on a high elevation south of the city, and affording a pressure of 450 pounds.

The Wallace light & water works

As early as 1889 a company was organized in Wallace for the purpose of exploring a system of light and water works. Some progress had been made in the enterprise, but the fire of 1890 consumed the rather rudimentary plant, and seriously embarrassed the new organization. The franchise and such property as remained were secured by E. D. Carter, who rebuilt the lines and began to develop the system.

In 1897 the Wallace Light & Water Company bought the interest of Mr. Carter. This company comprised J. A. Finch, A. B. Campbell, F. White, Richard Wilson and F. F. Johnson. At present the company is officered as follows: F. F. Johnson, president; R. E. Strahorn, of Spokane, vice-president and manager; E. J. Dyer, secretary and treasurer, and D. C. McKissick, local superintendent. The capital stock is $125,000.

Wallace hospitals

Quoting from the Murray Sun, of March 25, 1893: "The Providence Hospital at Wallace is an institution which has no equal of its kind in the state of Idaho, and no superior of its size in the United States."

In May, 1891, a temporary Miners' Union hospital was located in the American House building on East Bank street. Its financial condition was excellent, each miner connected with the various mines, with the Bunker Hill & Sullivan, having agreed to pay into the hospital treasury monthly dues of one dollar each.

But at that period, plans for a more elaborate institution were being prepared. On June 6, 1891, a committee of the Miners' Union and a citizens' committee met with Sisters Joseph and Madelaine, of Montreal, in the parlor of the Carter House to consider the question of the proposed new hospital. The citizens' committee comprised Messrs. McKissick, Gibson and Aulbach. They were asked by representatives of the Miners' Union if there was any valid objection to a transference by the miners of the hospital scheme to the sisters. There being none, the plan was consummated, which provided that the Sisters expend the sum of $30,000 in the erection of a handsome, four-story brick building, with Mansard roof and a frontage of 100 feet, not including the verandas. It was provided that the depth of the building should be from forty to eighty feet, the basement to be of stone, ten feet high. It was estimated that the structure would require 250,000 brick. Ground was broken for the new institution July 9, 1891.

Concerning this handsome structure the Murray Sun of March 25, 1893, said: "The origin of this hospital was with the miners' unions. It had become an imperative necessity. Sickness was prevalent, accidents numerous, and there was no place to take proper care of the unfortunates under their direction. The large membership of the unions justified the hospital, and in 1891 it was put under way. The generous offer of the people of Wallace was accepted, and the site located.

This created serious opposition in Wardner, which town also wanted the hospital, and the friction between the factions led to the outbreak of the Coeur d'Alene strike in 1892. The miners, however; ignored the Wardner feud; and a temporary hospital was established in Wallace and arrangements made to erect a $10,000 building, this being the agreement made with the citizens of Wallace. The miners; through the conduct of the temporary hospital, for seven months, found that it was almost impossible for them to make a success of it.

Several meetings of the executive committee were held and it was seriously proposed to give up the entire project and let each union take care of its own unfortunates. Then it was decided to enlist the aid of the sisters of mercy, and accordingly three sisters; including Sister Joseph, the present Lady Superior of the hospital; came to the Coeur d'Alenes on a tour of inspection and decided to accept the offer. The citizens committee of Wallace transferred the agreement to the sisters. At first only a temporary building was erected, a large, two-story frame structure. Finally, after numerous vexatious delays, work was commenced on the new brick building; pledged to cost $35,000, and this is now a reality. The citizens of Wallace have turned over a deed for the ground and water privileges. The building is located on a block in the eastern portion of Wallace. It is substantially built of brick; with granite foundation. It is practically four stories high; as the basement is lofty and as pleasant as any of the upper three stories. All the stories are hard finished and divided into large wards and single rooms; which are as cheerful as it is possible to make them. The hallways are broad and stair-way; of easy grade, and a hydraulic elevator is always in operation. Every modern convenience has been introduced, including hot air furnaces, dumb waiters, electric lights, etc. The capacity of the hospital is about 125 patients. The hospital was only put in complete running order last week, although occupied for nearly a year.

The first railroads

On Saturday, September 10, 1887, the first railroad to reach Wallace was completed to this point. It was a narrow gauge line, exploited by D. C. Corbin and associates, of New York city. At that time the Burke extension was contemplated, but right of way had not been secured. A depot, 24 x 80 feet, had been constructed, and regular trains were running to Wallace on September 19, 1889. This road was subsequently sold to the Northern Pacific company, washed out and abandoned in 1890.
The Northern Pacific Railroad Company ran

its initial train from Missoula to Wallace, in August, 1891. In 1900 a new round house, with a capacity sufficient to accommodate six engines, replaced an inferior structure. December 29, 1901, the company's officials first occupied the new depot, an elegant brick and concrete edifice, ornate and picturesque; located on Sixth street; on the north bank of the South Fork of the Coeur d'Alene river. The concrete, of which the greater portion of this building is constructed, is composed of "tailings" from ore concentrators, and cement. Its cost was between $8,000 and $10,000. A new addition to this building is contemplated.

On December 9, 1889, the Oregon Railroad & Navigation Company brought its first train in from Tekoa to Wallace, the terminus of the Tekoa division. At present, however, O. R. & N. business is carried as far east as Mullan, by special arrangement with the Northern Pacific Company. Both the Oregon Railroad Navigation Company and the Northern Pacific Company have standard gauge extensions up Canyon creek to Burke, a distance of seven miles, The new depot of the O. R. & N. was built in 1901. G. A. Newell, the present local agent at Wallace, has been with the company fourteen years, coming here June 11, 1889.

The first banks

The banking history of Wallace is marked by conservatism and business sagacity fully equal to that of any other town in Idaho, and superior to many. Unsettled financial conditions in 1893-4, of course, reacted upon all the banking centers of the Coeur d'Alenes, but in Wallace, particularly, recovery was rapid and financial loss far below the average, On January 2, 1891, articles of incorporation of the Coeur d'Alene Bank, of Wallace, were filed in the office of the recorder of Shoshone county, by virtue of which the institution was authorized to transact a general banking business in the state of Idaho, The capital stock was $50,000, the directors John A. Finch, Amos B. Campbell, Patrick Clark, Charles Hall and Joseph K. Clark, In 1893 the bank appears to have passed into the hands of George B. McAulay and Van B. DeLashmutt, who owned, also, the Miners' Exchange, of Wardner.

In April, 1893, the Bank of Coeur d'Alene asked for a receiver, attributing the cause of failure to bad debts, universal hard times and closing down of important mines. The liabilities were $70,679.73, of which $19,014.67 was due depositors, $18,561.25 to creditors holding certificates of deposit, and the remainder to outside state banks, including $42,429.07, overdraft on Bank of Wardner. In the Bank of Coeur d'Alene, Shoshone county had on deposit $18,435.22. This was secured by attaching the bank building in Wallace. Assets were, personal property, including notes, loans, discounts, etc,, $7229.20, and bank building at Wallace, $20,153.92.

In December, 1890, the Bank of Wallace had closed its doors, subsequent to a run which, for the time being, had been successfully withstood.

August 8, 1892, the First National Bank of Wallace was organized with a capital of $50,000. The officers were F. H., Johnson, president; Henry White, vice-president, Horace M. Davenport, cashier; Charles W. O'Neil, R, R, Neill, Richard Wilson, Albert Johnson, Henry White, C. E. Bender, and F. F. Johnson, directors. In 1903 M. J. Flohr succeeded Horace M. Davenport as cashier, The capital stock is $50,000; surplus fund, $10,000; undivided profits, $5,148 and circulation $42,700. President Johnson is treasurer of Shoshone county, president of the Wallace Light & Water Company and cashier of the Bank of North Idaho, at Murray.

The State Bank of Commerce, successor to the Bank of Commerce, which was organized in 1901, came into existence May 1, 1903. It is of-

ficered by Bennett F. O'Neil, president, Maurice H. Hare, cashier, Thomas L. Greenough, vice-president, and Charles Z, Seelig, assistant cashier. The directors are Thomas L. Greenough, Albert Burch, Ewen McIntosh, August Paulsen, G. Scott Anderson, Maurice H, Hare and Bennett F. O'Neil.

The Masonic Temple

Erection of the Masonic Temple, in 1896, was an event worthy of the originators of the project, and the result creditable to the city of Wallace. Shoshone Lodge, No. 25, A, F. & A, M., appointed a committee in May, of that year, to arrange for the building of a combination temple and opera house. The resultant organization was known as the "Masonic Building Association, Ltd." Estimated cost of the structure was $7,000, half of which amount was pledged by members of the association. For the remainder bonds of the denomination of $25, drawing six per cent interest, due in ten years, were issued. The committee in charge of building operations were, A. B. Campbell, chairman, F. F. Johnson, secretary, George Steward, E. H, Moffitt and L, W. Hutton.

January 1, 1897, the Temple was dedicated with appropriate Masonic ceremonies by members of Shoshone Lodge and chapter, O. E, S,, and several officers of the grand lodge of the state of Idaho. Spokane talent, mainly, was employed in the construction of this imposing edifice. The building was erected by Huber & Huelter; plastering by John Coleman; the heating apparatus was provided by the Griffiths Heating Company; the painting and interior decorations were the work of John McFarlane; the scenery was painted by Herman Ludke, while the stage mechanism was under direction of F. Thompson. Electrical appliances were provided by E, C. Morrow, of Wallace; Prusse & Zittel, of Spokane, were the architects. Total cost of the building was $20;000. The seating capacity of the opera house is 660; the stage is 58 x 28 x 45 feet in size. There are seven exits, toilet rooms, galleries and dressing rooms. The electrical plant consists of fourteen circuits. The second floor is divided into a lodge room, banquet hall, paraphernalia rooms and kitchen, Nearly all the fraternal societies of the city convene here and, at present, the hall is utilized as a court room. [Clarence Darrow argued a case here.]

Wallace becomes the County seat

During the year 1892 a county seat contest was sprung, ostensibly between Murray, Osburn and Wallace, but in reality between Murray, the county seat, and Osburn. Wallace threw the weight of its influence and votes in favor of Murray, and the contest proved unavailing. But in 1898, at the expiration of the six year limit, provided by law, to intervene between county seat imbroglios, the people of Wallace joined in a petition asking for removal of the capital of Shoshone County from Murray to Wallace. Practically there was no contest. Sentiment was universal that the county seat should be located on the South Fork of the Coeur d'Alene river, consequently Wallace was the only real contestant. Of 3,335 votes cast, Murray received 864; Wallace, 2,471.

The Public library

In December, 1902, the Wallace Public Library, near the corner of Sixth and Bank streets, was opened with pleasing social demonstrations. To the efforts of Rev. J. B, Ott, pastor of the Congregational church, the foundation and success of this institution are due, and he, at present donates his services as librarian. At its inception he paid the first month's rent, $30. On going to a coal dealer he was informed that the dealers in the city would undertake to heat the building gratuitously. Thus it was with the electric light company, and Mayor Connor's suggestion *that Mr. Orr apply for aid in paying the rent met with a cheerful response from the city council. Contributions of books flowed in, Mr. Orr taking the initiative with a liberal donation of volumes, and he was followed by the Episcopalians, who placed a generous addition in the city library. A free traveling library is sent out from the parent institution. No fees are charged for the use of books. Two committees are assisting in this commendable work, the male members being representatives of fraternal societies in the city, viz: George Warren, Masons; Otto Freeman, Odd Fellows; William Wourms, Woodmen of the World; Al. Crawford, Knights of Pythias ; J. R. Sovereign, William Stoll, Socialists; Robert A. Marshall, Order of Washington; William Adamie, Shoshone Club. The ladies' committee comprise Mrs. W. W. Wood, chairman; Mrs. A. R. Carpenter, treasurer; Miss Carrie Sovereign, Mrs. Harry Wood, Mrs. R. E. Seysler, Miss C. M. Hathaway, Miss Mamie Turner, and Miss Agnes Sutherland. The library contains over 1,000 volumes, and the patronage of Wallace and surrounding country is increasing.*

Idaho National Guard

In August, 1891, Company A, Idaho National Guard, a Wallace organization, was mustered into service by Captain Langdon, of Company C, Moscow. The company comprised of forty-three members, officered as follows: Captain, Thomas A. Linn ; first lieutenant, Robert Short; second lieutenant, E. G. Arment ; orderly sergeant, A. D. Short; second sergeant, William Hood; third sergeant, A. G. Larsen ; fourth sergeant, A. H. Utley ; fifth sergeant, Fred S. Bubb ; first corporal, Robert Duncan ; second corporal, Hugh Ross; third corporal, Jacob Lockman; fourth corporal, John Van Dorn. Musicians are, William Fitzpatrick, W. B. McCrary.

The Wallace Board of Trade

In response to a call of the citizens. signed P. F. Smith, chairman, a meeting was held at Masonic Temple, July 1, 1902, at which was organized the Wallace board of trade. The following officers were chosen, who, with the membership, comprise the only board of trade in the Coeur d'Alenes : P. F. Smith, president; O. D. Jones first vice-president; T. D. Connor, second vice-president; Herman J. Rossi, secretary ; L. L. Sweet, treasurer; executive committee - T. D. Connor, W. Hart, L. L. Odell, J. R. Sovereign, J. A. Allen, H. E. Howes, M. M. Taylor, M. J. Flohr, Theo. Jameson, H. J. Read, O. D. Jones, Harry White, George S. Warren, Jacob Lockman, W. A. Jones. The board has a membership of ninety, meeting the' first Tuesday in each month at the city hall. Harold J. Reach, chairman, P. F. Smith and Herman J. Rossi comprise a special committee on mining. Among future exploitations contemplated by the board are a street railway for Wallace, and electric lines from Wardner to both Burke and Mullan, requiring forty miles of rails, together with a wagon road, twenty-three miles in length, to tap the St. Joe timber belt, extending up Placer creek, over the divide, directly south of Wallace. The highest grade, near Summit, is ten per cent. In this vicinity several claim owners are now engaged in logging. The board has already constructed six miles of this road at an expense of $3,000. Annual dues of members are $10.

Churches in Wallace

The primary church organization in Wallace is the Episcopalian, the South Methodists coming next, who in turn, were followed by the Methodist Episcopalians. Subsequently the Baptists purchased the building of the South Methodists.

The Congregational church, of which Rev. J. B. Orr is pastor, is a thriving organization of five years' growth. Previous to 1898 Rev. Jonathan Edwards, at that time of Wardner, came occasionally to Wallace and preached to the Congregationalists, as did, also, Rev. Samuel

Green, state Sunday school superintendent of Washington. At present the Congregationalists, are holding services in Masonic Temple and other halls, but in August they will occupy a new church building, now in process of erection, corner of Fourth and Cedar streets, costing $3,500, exclusive of the lots. One lot is reserved, adjoining the new church, on which will be erected a gymnasium and bath rooms. The Congregational Sunday school numbers seventy-five pupils. On alternate Sundays Rev. Orr preaches at Kellogg.

Organization of the Methodist Episcopal church dates from September, 1894, when Rev. J. W. Craig, appointed to Wardner, Wallace, and at intervals, to Murray, Kingston, Gem and Burke, held services in this city. Rev. Craig, also, organized the local Epworth League. Rev. W. H. Selkirk came in 1895, succeeding Rev. Craig, from Pendleton, Oregon. He was appointed by Bishop Bowman, the annual conference sitting at Spokane. Rev. Selkirk was returned the second year, and was succeeded by Rev. R. W. Moore, Jr., and in turn was succeeded by Rev. M. R. Brown, in 1899. The latter remained in the Wallace field until 1902, when Rev. H. M. Hobbs, appointed by Bishop Earl Cranston, arrived in Wallace, September 14, and is the present pastor of the M. E. congregation. The organization has a pretty and commodious church building on the corner of Fourth and Pine streets, one block north from the new Congregational building.

"The Catholic church of Wallace, now in charge of Father Becker, assisted by Father Beusman, was dedicated by Bishop Glorieux, Sunday, October 20, 1896. This was the second church organization in the county, of that denomination. The present church edifice, corner of Pine and Second streets, was built under direction of Father Keyzer, who was the first clergyman in charge. Father Becker, present incumbent, came to Wallace, May 11, 1899. The labors of Fathers Becker and Beusman are not confined to Wallace. They visit the outlying missions of Wardner, Mullan and Burke. The church organization of Wallace comprises fifty families.

"The Baptist denomination has a regularly organized church in Wallace, holding services in what was once the Methodist Episcopal church, on Bank street. In 1896 it was in the charge of Rev. Lewis Smith, of Spokane, assisted by Rev. A. M. Allyn, at that period district missionary for Washington and northern Idaho. The church edifice, corner of Cedar and Fourth street, as well as others throughout the county, were built under direction of Bishop Talbot.

"In March, 1901 a new brewery, and the only one at present in Wallace, was thrown open to the public. The three principal buildings were erected in 1900, solid, substantial brick structures, located east of the Carter House, at the termination of the O. R. & N. Company's warehouse tracks. The office is a single story brick building, 27x62 feet in size. The brewery, a twin building, has a frontage of fifty-six and a depth of eighty-six feet, and is four stories high. It has an annual capacity of 80,000 pounds of malt, rice and hops and, with the adjoining bottling plant, cost $50,000. It is the property of the Sunset Brewing Company (Incorporated), and under control of the following board: David Holzmann, president; Jacob Lockman, secretary, treasurer and manager; I. Henry Beckman, Freida Lockman and Joseph A. Rubens."

Chapter VI

Downtown Wallace From The Late 1880s to Mid 1930s

SIXTH STREET, WALLACE.

Sixth Street, Wallace, circa 1890. *(Courtesy Butch Jacobson)*

Holley, Mason, Marks & Co., built in 1890, and located on Bank Street. *(Barnard Stockbridge photo, courtesy Butch Jacobson)*

The Jones and Deane Building, circa 1892. This structure burned in 1933. *(Barnard Stockbridge photo, courtesy Butch Jacobson)*

Federal troops marching down 6th street, during the labor troubles in 1892. On July 14, 1892, the entire mining district was placed under martial law after union miners blew up the Frisco mill in Burke Canyon. The rampaging miners did not resist the troops and hundreds were taken into custody. Since no jail in the district could hold all the prisoners, two cottages and a large warehouse served as a temporary jail in Wallace to hold them for a later trial. Two non-union and three union men were killed and others wounded by company guards in the exchange of gunfire. *(Courtesy Butch Jacobson)*

Wallace, looking east on Bank Street, during the 1892 labor dispute. The encampment in the foreground, at approximately Bank and King Street, was for the military troops called in after the Frisco Mill was dynamited. Mallon's brewery and ice pond are to the left. *(Photo Courtesy Butch Jacobson)*)

Another military encampment in Wallace during the 1892 mining labor disputes. View is looking west over downtown Wallace. Note how rapidly the town rebuilt after the 1890 fire. *(Photo Courtesy Butch Jacobson)*)

Wallace, Idaho, view of 6th Street, circa 1895. *(Barnard Stockbridge photo, courtesy Butch Jacobson)*

Wallace, Idaho, Sixth Street taken from Bank Street, circa 1898. *(Barnard Stockbridge photo, courtesy Butch Jacobson)*

The Masonic Opera Hall was built in 1896. This building escaped the 1910 fire but was later destroyed by fire in 1916. It had a seating capacity of 700. The Clarence Darrow defense of Steve Adams took place on the top floor of this structure.

Many of the old stars of the "Golden Age" of theatre preformed in the Wallace Opera House. During its time, the city was noted as having the best shows in town between Butte and Spokane.

When a show came into the Opera House a special train was made up to bring residents of the towns of Kellogg, Burke, Mullan, and other points between, to Wallace to attend the openings.

The Masonic Lodge rooms occupied the upper floor. This famous landmark was located on the site of the present City Swimming Pool. *(Barnard Stockbridge photo, courtesy Butch Jacobson)*

The Delashmut Building, Murphy/Lucas Clothing Company occupied one floor. The old court rooms are on the second floor, circa 1910. *(Barnard Stockbridge photo, courtesy Butch Jacobson)*

Interior Dick Daxon's Saloon, circa 1890. *(Barnard Stockbridge photo, courtesy Butch Jacobson)*

The Gem Restaurant was on Cedar Street, adjacent to the Arment Building, circa 1910. *(Courtesy Butch Jacobson)*

This photo is hanging in the Shoshone County Courthouse. It is a typical scene of migrant workers as they posed for a photo along the old Mullan Trail. It wasn't uncommon for entire families to travel that route. This photo appears to have been taken in or near Wallace, circa 1890s. *(Photo Courtesy Shoshone County Courthouse, Barnard/Stockbridge photo)*

Jameson's Billiard Hall in the Jameson Hotel at 304 Sixth Street, circa 1906. *(Courtesy Butch Jacobson)*

Wallace, Idaho, circa 1894. *(Barnard Stockbridge photo, courtesy Butch Jacobson)*

The Steward Brothers Drug Store at the corner of 6th Street, circa 1895. *(Barnard Stockbridge photo, courtesy Butch Jacobson)*

The Hall of the Wallace Fire Department taken on April 30, 1902. The Good Bye Banner was placed there to commorate the dismantling of the old fire station to be replaced with a new one. *(Courtesy Butch Jacobson)*

Placer Creek flood in 1897 at Wallace, looking up King Street. (Photo Courtesy Butch Jacobson)

Interior of Fahle's Saloon in Wallace, circa 1900. *(Barnard Stockbridge photo, courtesy Butch Jacobson)*

Sweet's Bar. Lew Sweet is on right behind bar. *(Barnard Stockbridge photo, courtesy Butch Jacobson)*

The Stanley Hotel on Bank Street, circa 1900. *(Courtesy Butch Jacobson)*

The Following Information Was Provided by Phillip Christman

The Stanleys originally operated the Tiger-Poorman boarding house in Burke. They then came to Wallace, making their home along Placer Creek at what is now the site of Pine and 2nd streets. Mrs. Stanley was the owner of the Stanley Hotel and Missoula House, both of which were destroyed in the 1910 fire.

Following a short residence in St. Paul to care for her aged mother, she returned to Wallace and erected the present Stanley Hotel on Bank Street.

Both Augusta and Charles Stanley had Germanic and Polish roots. Charles reportedly was born in Frankfort but his mother died in Poland and his full sister lived in Poland as well. His birth surname appears to have been Berndt but at some point he changed it to Stanley. His half brother was Maxamillian Toltz who was a partner in the large Minneapolis Engineering firm of Toltz, King, and Day. That firm designed many of the large bridges and buildings in Minneapolis and is still a going concern today although under a changed name.

Prior to forming this firm Max designed roads in Germany and helped design tunnels and bridges for the Great Northern RR under owner Hill.

It is unknown what drew the Stanleys to Burke and Wallace but the area was growing and the silver mines probably offered an attraction.

After Charles died in 1905, Augusta continued to run boarding houses and hotels with her son Charles.

She invested heavily in a lot of the new mine stock offerings and had a chest full of mining stock certificates when she died. Unfortunately the vast majority of them were worthless.

The elder Stanleys had 5 children, 4 boys and 1 girl. The eldest, Charles, helped his mother run the hotel but apparently died at age 41 after ingesting moonshine made in a creosote barrel from the RR. The 2nd child, Ella Evie Stanley died at age 2, prior to the Stanleys leaving Minnesota. The 3rd child Rosa married well but divorced and her 3 subsequent marriages were not happy. The 4th child Art became very successful and was influential in the newspaper industry in California. The last child, Alvin, remained in the Wallace-Avery area and worked mainly as a house painter. He was married twice, the first marriage resulting in a child named Roger Stephan Stanley. The second marriage produced Alena (Phillip A. Christman's mother) and Philip Stanley. Roger sired 7 children with 2 wives while Philip had 2 children. Phillip was an only child until his parents adopted a Korean orphan.

Roger's family still is an active part of the Wallace community. Phil Stanley was superintendent of the Avery school district for many years.

The below information was taken from the Wallace paper obituary (unknown date) on Augusta-probably written by one of her surviving children:

The "Stanley Furnished Rooms" building or boarding house. It is unknown if this was located in Burke or Wallace. The photo and information was provided by Phillip Christman, a relative of the Stanleys.

WALLACE OBITUARY:

Mrs. Augusta Stanley, 90, one of Wallace's oldest residents, both in age and duration of residence, is dead. The end came in a Wallace Hospital at 7 o'clock Saturday evening following a brief illness.

At her bedside were a daughter, Mrs. Rose McDonald – Vancouver, B .C, Arthur F. Stanley, Los Angeles, and Alvin W. Stanley, Wallace, Idaho.

She also leaves eight grandchildren and 15 great grandchildren. Mrs. Stanley, had been active and in good health almost to the end. Mrs. Stanley came to the Coeur d'Alenes in 1888 and had resided in this area continuously since that time. She was born in Milwaukee, Wisconsin on 10/29/1862, and spent her girlhood in St. Paul. She married Charles Stanley there and came to Burke in February, 1888. Mr. Stanley died here in 1905.

Cedar Street, Wallace, circa 1907. *(Barnard Stockbridge photo, courtesy Butch Jacobson)*

The Hathaway House, Bank and Sixth streets, Wallace, circa 1906. *(Barnard Stockbridge photo, courtesy Butch Jacobson)*

Wallace Public Library

Lucie Wallace started a library shortly after she arrived. In December, 1902, the second Wallace Public Library, near the corner of Sixth and Bank Streets, was opened by the effort of Rev. J. B. Orr, pastor of the Congregational Church – Carnegie's legacy.

The Carnegie Library opened in 1911

Andrew Carnegie may have been the richest American of all time. The Scottish immigrant sold his company, U.S. Steel, to J.P. Morgan for $480 million in 1901. That sum equates to about slightly over 2.1% of U.S. GDP at the time, giving Carnegie economic power equivalent to $372 billion in todays market.

Carnegie used his fortune to establish Carnegie Hall, and founded the Carnegie Corporation of New York, the Carnegie Endowment for International Peace, the Carnegie Hero Fund, the Carnegie Institution for Science, the Carnegie Trust for the Universities of Scotland, Carnegie Mellon University, and the Carnegie Museums of Pittsburgh, as well as 2,509 libraries across the United States and the United Kingdom.

Steel magnate Andrew Carnegie became one of the wealthiest industrialists in America but gave much of his fortune away for the "improvement of mankind." When asked about the best philanthropic gift he could give to a community, his answer was a free library. Between 1886 and 1919, Carnegie's donations of more than $40 million paid for 1,679 new library buildings in communities large and small across America. Many still serve as civic centers, continuing in their original roles or fulfilling new ones as museums, offices, or restaurants.

The Carnegie Library. *(Courtesy Butch Jacobson collection)*

Andrew Carnegie.
(Public domain)

In his autobiography, Carnegie remembered that, as a child, "I resolved, if wealth ever came to me, that it should be used to establish free libraries." And he did, providing public libraries to communities across the country, all engraved, at his request, with an image of a rising sun and "Let there be light."The patron of these libraries stands out in the history of philanthropy. Carnegie was exceptional in part because of the scale of his contributions. He gave away $350 million, nearly 90 percent of the fortune he accumulated through the railroad and steel industries. Carnegie was also unusual because he supported such a variety of charities. His philanthropies included a Simplified Spelling Board, a fund that built 7,000 church organs, the Carnegie Institute in Pittsburgh, the Carnegie Foundation for the Advancement of Teaching, and the Carnegie Endowment for Peace. Carnegie also stood out because some questioned his motivations for constructing libraries and criticized the methods he used to make the fortune that supported his gifts.

Carnegie libraries in Idaho, where 11 libraries were built from 11 grants (totaling over $138,000) awarded by the Carnegie Corporation of New York from 1903 to 1914. Construction was delayed due to the 1910 forest fire that broke out August 20. The fire eventually burned three million acres of timber, and 85 were killed. Rains on August 31 finally ended the fire. The library was eventually built in April 1911 at a total cost of $15,300.

The Nine Mile Stage in front of the Wallace Cafe. Note the sleigh runners. Back in these times, Wallace had harsher winters and received greater levels of snowfall than is typical of the present. Snow removal was accomplished manually or with the aid of horses, circa 1889. *(Courtesy Butch Jacobson)*

The bar in Fahle's Hotel located in the Furst Building, built in 1900, at 517 Cedar Street. Their letter-head read: "Fahle's Hotel, Headquarters for Commercial and Mining Men." The safe has the building owner's name, John G. Furst, across the front. Neither calendars on the wall – Fahle's own with the woman's picture and Wallace's First National Bank with "Mon. 16"– divulge the year of the photo. When built, John Furst named his hotel the Centennial Hotel. It later became Fahles Hotel when Bill Fahle bought it.*(Courtesy Butch Jacobson)*

Levi "Al" and May Hutton, circa 1888.. *(Courtesy Butch Jacobson)*

During a rock drilling contest on July 4, 1907 at Wallace, Idaho, George Baker and Walter Joy of Murray, beat C. M. Paterson and Pete Haff of Osburn, who were the former holders of the DCA drilling record. The prize was $280 to be split between the two of them. *(Courtesy Butch Jacobson)*

Miner's Home Saloon, circa 1906. *(Barnard Stockbridge photo, courtesy Butch Jacobson)*

Sweet's Hotel and Jameson's in Wallace, built in 1907, photo circa 1907. *(Barnard Stockbridge photo, courtesy Butch Jacobson)*

The November 1906 Wallace flood. This flood washed out railroad tracks and telephone and telegraph lines, leaving Wallace cut off from the rest of the world. Sewers backed up and drinking water was contaminated. Worse yet, receding floodwaters left four inches with alleged toxic lead silt behind.
(Photo Courtesy Butch Jacobson)

The November 1906 Wallace flood. Upper photo is looking northwest on 6th Street. The lower photo is looking south on 6th Street. *(Photo Courtesy Butch Jacobson)*

The interior of the Miner's Home Saloon, Wallace, circa 1906. *(Courtesy Butch Jacobson)*

The lobby of the Samuels Hotel in 1907. The Victorian-style hotel at the corner of Cedar and Seventh streets was considered one of the finest hotels in the Northwest when it was built in 1907. It was razed in 1974, and has the unfortunate distinction of being the first and only brick building to be razed in downtown Wallace. *(Courtesy Butch Jacobson)*

Western Bar and restaurant. Typical of small frontier towns, saloons, and taverns outnumbered other types of business establishments. Note the circular sign over the head of the man in the middle. It was advertising beer from Wallace's Sunset Brewery. During the 1910 fire, the brewery went up in flames and 2,000 barrels of foaming suds poured into the streets. *(Courtesy Butch Jacobson)*

The Wallace fire department horse and ladder wagon, circa 1907. *(Barnard Stockbridge photo, courtesy Butch Jacobson)*

The Wallace Fire Department, circa 1915. *(Barnard Stockbridge photo, courtesy Butch Jacobson)*

An article appeared in the Idaho press, February 3, 1910, which I paraphrased:

Ski Jumping in 1910, on fifth Street in Wallace Idaho. John Rubke, an employee of the post office, often entertains crowds of people during the noon hour by exhibiting his ski jumping technique on the hill at the west side of the post office block.

Rubke goes to the high cribbing above Fifth Street for his start and travels more than 100 yards before he reaches the bump where he makes his leap. He strikes this at a high rate of speed, leaps in the air for more than 40 feet, and lands part way down the hill.

Rubke became a devotee of the sport while a resident of Red Wing, Minnesota, one of the centers of ski jumping in the United States. B. O. Skonnard, another ski jumper, who watched Rubke's performance yesterday, and who has participated in the sport in the old country, said that Rubke handled himself well, but the skis were not as smooth as those in the old country. Rubke skis were not burnt in tar and otherwise treated until they were as smooth as glass. If his skis had been treated that way, he could have attained a much greater speed.

Another article concerning Rubke appeared in the *Wardner News*, on February 12, 1910:

John Rubke, for a number of years a miner living here in Wardner has set a record for the Northwest in the matter of ski jumping in a test in Wallace. Rubke cleared the ground for distance of 60 feet. This is a record hard to equal even by the experts of the northern states. Rubke comes from Minnesota where the artist practices.

(This photo is from the collection of Kathryn E. Eichwald, she was a longtime secretary to Henry "Hank" Day, president of Day Mines. She allowed Butch Jacobson to copy this and other photographs from her collection. Courtesy Butch Jacobson collection)

The ski hill John Rubke jumped on Fifth Street in 1910. The building on the left is the North Idaho Trading Company, owned by Rich Asher. This location is 5th and Bank Street – H. F. Magnuson Blvd., May 2012. *(Bamonte photo)*

Elk's Club, built in 1924-25. *(Barnard Stockbridge photo, courtesy Butch Jacobson)*

The Hale Building. The Fremont cafe occupies the bottom floor, circa 1930. *(Barnard Stockbridge photo, courtesy Butch Jacobson)*

The Wallace High School, built in 1933, was located at the northeast corner of River and 3rd streets. *(Courtesy Butch Jacobson)*

Runaway ore cars from Nine Mile Canyon loaded with zinc concentrates crashed in front of the Northern Pacific Railroad Depot, fortunately missing the depot itself, circa 1913.

The chateau-style depot on the north bank of the South Fork of the Coeur d'Alene River was completed in May 1902. The concrete used in the construction of this building contained tailings from local concentrator mills.

The depot has since been moved to the south side of the river and is now a railroad museum. The long adjoining building was the original narrow-gauge railroad depot and moved to this location to serve as the freight house. It survived until 1976. The NP House hotel is visible across the river. *(Photo Courtesy Butch Jacobson)*)

The Wallace Fire Department in 1907. The horses were Punch & Judy. It was with this equipment that the local department fought the fire that destroyed so much of Wallace in 1910. *(Courtesy Butch Jacobson)*

The Wallace Fire Department, circa 1925. *(Courtesy Butch Jacobson)*

The Northern Pacific Depot before and during the move. In 1986, the Northern Pacific Railroad Depot was moved 200 feet across the south fork of the Coeur d'Alene River to make room for two massive highway columns. The inset in the lower photo is a staged photo of Heather Branstetter's grandfather, Ken Branstetter, who worked at the N. P. Depot for years. *(Courtesy Butch Jacobson)*

A Brooks Bakery delivery wagon, circa 1900. *(Courtesy Butch Jacobson)*

A rail line snow plow used for the heavy snowfalls, circa 1900. *(Courtesy Butch Jacobson)*

The ambulance on River Street, across from ball park, circa 1913. *(Courtesy Butch Jacobson)*

The Northern Pacific Restaurant in Wallace on Sixth Street, circa 1910. *(Courtesy Butch Jacobson, Barnard/Stockbridge photo)*

Cedar Street looking west, circa 1915. *(Courtesy Butch Jacobson)*

The Northern Pacific passenger train arriving at Wallace from Spokane, circa 1920. *(Courtesy Butch Jacobson)*

A northern Pacific engine parked at Wallace with the crew posing, circa 1920. *(Courtesy Butch Jacobson)*

The 1933 Placer Creek Flood. *(Courtesy Butch Jacobson)*

Looking northwest down King Street during the 1933 Placer Creek Flood, December 23, 1933. *(Courtesy Butch Jacobson)*

The old Wallace swimming pool up Placer Creek. It washed out in the 1933 Placer Creek flood. *(Courtesy Butch Jacobson)*

Dr. Charles Mowery courting his future wife, Doris, at their home in Wallace on July 10, 1902. Doris and Charles are on the porch to the far right. *(Courtesy Don Harvey husband of Betty Mowery, daughter of Charles and Doris Mowery)*

Tourist camp on the outskirts of Wallace, circa 1915. *(Courtesy Don Harvey husband of Betty Mowery, daughter of Charles and Doris Mowery)*

Anderson Hardware Company in Wallace, Idaho.*(Courtesy Anderson collection)*

The interior of Anderson's Hardware Store in Wallace, Idaho, circa 1910. Some of the things shown in the store include: guns, fishing poles, fishing creels, rakes, hammers, lawn mowers, axes, children's wagons, shovels, hatchets, ammunition, buckets, oil cans, chisels, door hardware, hinges, picks, brooms, shovels, pliers, drill bits, hand saws, etc. This store was located on Bank Street just east of where the Wallace Mining Museum now sits. *(Courtesy Anderson collection)*

Anderson Hardware warehouse in Wallace, Idaho, circa *(Courtesy Anderson collection)*

East Wallace before 1910 fire. *(Courtesy Anderson collection)*

Parade in Wallace going down Cedar Street on July 4th, 1921. *(Courtesy Anderson collection)*

Placer Creek as it runs into Wallace, Idaho, in 1922. *(Courtesy Anderson collection)*

Wallace Train Depot, September 10, 1887. *(Courtesy Anderson collection)*

Elks Round Up Parade on Bank Street in Wallace, on June 29, 1922. *(Courtesy Anderson collection)*

The hillside of Wallace in 1920, in the vicinity of Irving Anderson's home. *(Courtesy Anderson collection)*

The Wallace Gun Club in February 1922. *(Courtesy Anderson collection)*

The ALB Circus came to Wallace, circa 1922. *(Courtesy Anderson collection)*

The Wallace Burke Stage across from the Barnard Studio at the east end of Cedar Street in 1924. The man on the right is Guy Ghigleri. The following song, was alledgedly written by a Mr. Thearian

**From the banks of the CDA river
to the mighty Glidden shore
From the downtown streets of Wallace
up to Otto Olsen's store**

**There's a bus of majestic splendor
She's known quite well by all
She's the modern accommodation
Called Ghigleri's Cannonball**

**Listen to the rattle, the rumble,
and the roar
As she wanders up the woodland by
Canyon creeks wild shore**

**Hear the mighty rush of engine hear
the rough old mucker's call
As we go on up the canyon
In Ghigleri's Cannonball**

**Great cities of importance
can be seen along the way
Woodland Park and Gemtown, and
Frisco so they say**

**Yellow Dog, Mace and Black Bear
and Burke above them all
can be reached by no one other
than Ghigleri's Cannonball**

Burke School, circa 1907. A.A. Amonson grew up in Burke and graduated in 1911 in a class of two. He worked as a bookkeeper for numerous mining companies and an insurance company. His son John, a fount of knowledge about the area's history and its mining activities, taught classes on the subject at Wallace High School and North Idaho College, and worked at the Wallace District Mining Museum. John's great uncle, Carl, discovered the Hummingbird Mine. The Hercules #5 and the Hummingbird were two distinct mines. The Hummingbird was across the gulch from the Hercules. *(Courtesy Wallace District Mining Museum)*

The mining town of Mace on Canyon Creek, June 1907. Mace was named for Amasa B. Campbell (inset) who was a principal owner of some of the most productive mining properties in the Coeur d'Alenes (including the Hecla, Standard, and the Gem). He built a home in Wallace around 1890, but subsequently moved his family to Spokane, where he built a mansion in the historic Browne's Addition. The home was donated to the Eastern Washington State Historical Society by his daughter Helen (Campbell) Powell. It is now part of the N. W. Museum of Arts a& Culture. *(Courtesy Spokane Public Library)*

Transporting mining equipment (a babbitt bearing crusher) up the Gorge Gulch Road, possibly to the Hercules #3. Gene Day's horse team. Ed Albin is on the far left of the photo. He was the Hercules teamster and a cousin of Guido Bardelli, circa 1902. *(Courtesy Spokane Public Library Northwest Room)*

Cottage Grove, east of Wallace, and south of the Providence Hospital. This was on the south side of Highway 10, photo taken 8-16-20. *(Butch Jacobson collection)*

The Ron D'Voo Tavern, east and past Cottage Grove next to the old Highway 10. It was an old log cabin night club from the mid 1930's to theearly years of the 1960s. It was destroyed for the constructin of the new interstate Freeay I-90. *(Butch Jacobson collection)*

The Ron D'Voo Club from the *Mullan News*, October 23, 1936 *(Butch Jacobson collection)*

In or near Wallace, circa 1890. *(Anderson Collection)*

Placer Creek, above Wallace, circa 1890. *(Anderson Collection)*

Chapter VII

Significant Places, People, and Events Related to Wallace

Placer Creek, above Wallace, circa 1890. *(Anderson Collection)*

History of the Shoshone County Courthouses

Pierce, Idaho, was the county seat of Shoshone County from 1862 to 1885. Although Pierce gained a large population for a year after gold was discovered there in 1860, most miners soon moved on to other camps. By 1880, Shoshone County had only 469 people left, of whom 296 were Chinese. A new gold rush, in 1883, 80 miles farther north (Eagle/Murray area) led to the removal of county government to Murray in 1885, Pierce's courthouse became a private home. The distance between Pierce and Murray is 211.7 miles.

Shoshone County's original courthouse and Idaho's earliest public building is located and still stands in Pierce, where it was built in 1862. It is one of twenty-six sites of the Nez Perce National Historical Park in North Central Idaho, and located one block off Main Street (Idaho Highway 11) in Pierce, Idaho.

It is located on a grassy lot on the main street of Pierce, and surrounded on three sides by residential development. The 20' x 40' log building with board and batten veneer was rehabilitated in 1990. Although its original context may no longer exist, it is situated among a number of other suitably scaled mining town structures. The building is open on request and on summer weekends. A wayside exhibit summarizes the impact of gold mining on the Nez Perce. Eight panels inside the courthouse detail the area's history. Mining led to the reduction of the Nez Perce Reservation through the treaty of 1863. *(Bamonte photo)*

The Shoshone County Courthouse when it was located in Murray, on the northeast corner of Main and Keeler. The old jail was located across the street. The courthouse building belonged to Steward Fuller. The county rented it from him for use as the courthouse. It was used as a courthouse from 1885 to 1899. The building later became the headquarters of the Coeur d'Alene Placer Mining Co., founded by Barry N. Hillard. From 1940, until it was abandoned in 1978, it was the Courthouse Tavern. In November 1978, the building was placed on the National Register of Historic Places. A movement to restore it was underway when, on February 21, 1997, it collapsed under the weight of heavy snow. With the help of a county insurance policy, it was completely reconstructed. *(Courtesy Butch Jacobson)*

Murray Loses the County Seat to Wallace

This information comes from the research my wife and I did in our book *The Coeur d'Alenes Gold Rush and Its Lasting Legacy*

By 1888, most of the population and the majority of the mining interests were centered along the South Fork of the Coeur d'Alene River and its gulches and tributaries; and traveling to the county seat at Murray was no small inconvenience. As a result, residents of both Wallace and Osburn began to agitate for the relocation of the county seat. Talk of a change from Murray began to surface in local newspapers as early as 1888.

Although a move from the North Side to a more central location along the South Fork made logical sense, the first major issue to arise was who had the authority to make such a decision. The October 20, 1888, *Wallace Free Press* reported that it was the jurisdiction of the territorial legislature, but others argued that applied only to a newly established county and a majority vote of the people could choose the location. Despite the hope of putting it to the vote in the upcoming November election, no such measure made it to the ballot. However, the advocates for the move were just getting warmed up, and the newspaper articles clearly revealed the rivalry and friction among the leaders of the communities involved.

Although it does not appear they were the originators, two of the main agitators behind this movement were William Clagett and Weldon Heyburn. They felt the location of Osburn would be the most fitting one, largely because it was centrally locat-

Senator Weldon B. Heyburn, circa 1910. *(Courtesy Library of Congress)*

ed and the width of the valley at that point provided ample room for a town to grow. Clagett and Heyburn were men of some influence, having established their reputations in the legal arena, by their present involvement in the territories movement toward statehood, and their participation in writing the new Idaho State Constitution. Furthermore, both were brilliant politicians.

They also had personal reasons to advocate for Osburn. By 1888, Clagett had a home built there, located on the Mullan Road, the main street of Osburn. He opened a law office across the road on the south. Although the town only had a population of about twenty male residents (the voting population), it was a logical choice for Clagett; it was at the base of the Two Mile Road, which then was the most direct route between Murray and the center of activity along the South Fork.

In October 1889, Heyburn purchased Lee George's two-story log building at the east end of the Osburn plat, formerly known as Georgetown, and had it converted into a comfortable residence and a law office. It was right next door to Clagett's home. (Weldon Heyburn's brother Elwood built on the other side of Clagett. Elwood's daughter later married Clagett's son, George.) Although Heyburn had a law office in Wallace, he had been involved, since 1884, in some mining ventures, including the Polaris, near the town of Osburn.

In March 1889, Clagett proposed appealing to the U.S. Congress for the right of voters of Shoshone County to locate the county seat. He and John M. Burke were approved as delegates to make the trip to Washington, D.C., and a committee was formed, with representatives from every town in the district, including Murray, to collect money to fund their trip. The audacity to ask Murray for money angered its residents, who were already disgruntled because of a personal affront to the town; in his argument for Osburn, Clagett knocked Murray as being "abandoned and deserted." (Ironically, at that time some promising lode discoveries were being made, giving Murray a new boost.) They also felt betrayed by Clagett, who had been a strong advocate for Murray and had great financial successes while residing there. Consequently, not only did the Murray residents keep a tight hold on their purse strings, they had a few unkind words for Clagett. (As a senator in Montana in 1872, Clagett wrote the Bill thata created Yellowstone National Park.)

By the fall of 1889, the race was heating up. Both Wallace and Osburn were putting forth a great deal of effort to wrestle the county seat away from Murray, which the latter town was not about to let go of without a fight of its own. The Osburn supporters characterized Wallace as nothing more than a cedar swamp squeezed into a narrow valley, where there was no room for expansion. The Wallace supporters pointed out that Osburn had nothing to offer by way of infrastructure or accommodations. Back and forth it went, each side hustling to gain additional supporters.

At this point, with their own vested interests at stake, Clagett and Heyburn stepped up their efforts with another angle. Shortly after the district court adjourned in January 1890, they began circulating petitions to have the next term held at Osburn. They quickly added a few hundred signatures to their petition, which would be presented to the Supreme Court of Idaho. Of course, when the Wallace supporters heard of this movement, they also began lobbying to have the court sessions held in Wallace. The argument presented to the Supreme Court was that, with one or two exceptions, all of the nearly one-hundred cases on the docket for the next session involved residents and issues from the South Side. It also pointed out the distance to Murray imposed inconveniences and expenses to all involved, further claiming the Murray courthouse was in an unsafe condition.

The logical place to hold court would naturally have been at Wallace, and its advocates were certain the court would rule in their favor. Not only had they gathered well over a thousand signatures, but Wallace was a town of some consideration. It was situated in the heart of the most populated region, had ample hotel accommodations and restaurants; a courtroom, jury rooms, and a jail, all of which were offered to the county free of charge. It also had the best transportation connections. It was highly unusual that an approval was given to hold the district court in a locale other than the county seat (at Murray), but many were shocked to learn, in February 1890, that Osburn had been granted the honor, despite not having a single facility in which to hold court or accommodate those involved. Apparently, as it turned out, Clagett and Heyburn had the ear of Judge Willis Sweet, the politically motivated district judge, who had the authority to make the final decision.

The approval for the new venue was only for the two 1890 sessions, the first of which was to commence on May 26th. To accommodate the district court, and in anticipation of giving the community of Osburn a foot in the door to obtain the county seat, Stephen V. "Bill" Osburn, for whom the town was named, built a courthouse and a hotel, which were ready just in time for the start of the court session. His new hotel had twenty rooms and housed about a hundred guests. On July 16, 1915, when, twenty-five years later, the courthouse was being demolished, the *Wallace Press-Times*, under the heading "Old Court House Is Being Wrecked,"gave the following description:

Osburn's Hotel,
Osburn, Idaho.
Finest hotel in the South Fork Valley. Accommodations first class in every respect. Large and well ventilated rooms, and a table service that is unsurpassed anywhere. Large stable connected with hotel.
WM. OSBURN, Proprietor.

An ad for Osburn's Hotel at Osburn, from the November 1, 1890, *Wallace Press*. It was built by Bill Osburn to accommodate those attending the First District Court sessions held in Osburn for one year (1890). It was known as the Osburn Inn for many years, but was called the Midway Inn at the time it was destroyed by fire on February 19, 1913.

To the casual passerby the old relic presents no picture but that of a ramshackle warehouse that may possibly have completed its day of usefulness many years before; but to the men who took part in the upbuilding of the greatest silver-lead district on earth, other things are called to mind – the stirring days when Idaho first entered the sisterhood of states, when this mining camp was involved in a death grapple with its neighbors for the county seat.

Although the district court returned to Murray the following year, where it continued to hold future sessions until the county seat was finally moved to Wallace in 1899, the act of holding court in Osburn for that one year created the mistaken notion that Osburn had been the county seat. It never was. Lifetime resident and former county prosecutor, Richard Magnuson, who was also the county's official historian, made repeated efforts to correct that myth, but to little avail. (Case in point: Shoshone County's website states the county seat moved to Osburn in 1890. It further states the county seat moved to Wallace in 1893, although that move was approved by the voters in 1898 and took place in 1899.)

Concurrent with the news, in the spring of 1890, regarding the district court's approved change in venue, was the announcement that a bill, H. R. 6475, had been submitted to the U.S. House of Representatives to give the voters of Shoshone County the right to determine the location of its county seat. The bill passed, and the measure was placed on the October 1st ballot. Although Osburn lost, Spokane's *Morning Spokesman* reported the results prematurely, announcing that it had won; this had the unfortunate effect of helping to cement the myth of Osburn having been the county seat. According to the November 1, 1890, *Wallace Press*, the results of the election were as follows: the total votes cast were 2,057, of which Osburn received 29, Wallace 706, Murray 364, Gem 1, Kellogg 3, and Kingston 1. (These figures differ slightly from those published in the same paper the day immediately following the election.) Had Wallace not been hit with a devastating fire in July, which leveled much of its downtown core, it may have stood a better chance of securing the county seat. Because no city received a majority of the total vote, the county seat was to remain at Murray, at least for the time being.

No time was wasted in resuming the removal efforts. Again, Wallace and Osburn were the main contenders and both actively courted the voters for support, especially those in the more heavily populated Wardner precinct west of Osburn. In February 1891, a bill was submitted to the Idaho State Legislature to place the issue on the ballot in the November 1892 general election. Idaho law mandated that towns interested in the county seat circulate petitions, and the one with the most signatures would be placed on the ballot as the contender against the present county seat. In accordance with the law, on the first of March 1892, those petitions were being put into circulation. The pressure was on to gain every possible vote, because the state of Idaho required a two-thirds majority, rather than the simple majority required by the U.S. Congress in the previous election.

Wallace was still the largest town on the South Fork, with even more to offer than during the earlier race, and was expected to be the victor. To show their serious intent, some of its ardent supporters bonded themselves to the county commissioners for $20,000, the purpose of which was to enable the city to donate a suitable piece of property and to build a two-story building for county purposes. The bond would also cover the cost of moving the county records from Murray to Wallace. Adam Aulbach also was throwing his support behind Wallace, where he had moved in July 1890 after leasing the *Coeur d'Alene Sun* to a syndicate of Murray businessmen. In August, he started the *Wallace Press*, which he used to praise the town's fine attributes. In May 1892, the new managers terminated their lease of the *Sun*, and for a couple of months, Aulbach searched unsuccessfully for someone else to run it. During this time, he was getting a lot of heat from the government and the military, whose wrath he incurred by his unconstrained journalistic support of the unions after angry miners had blown up the mill at the Frisco Mine in July. In response to the combined contentions, on July 16, 1892, Aulbach printed his final issue of the *Wallace Press* and returned to Murray to resurrect the *Sun*.
As election time neared, Wallace appeared to be in a solid lead, but as noted, the entire South Side district had become embroiled in the first of its contentious mining wars. Murray, on the other hand, was experiencing a period of renewed prosperity and stability. Its population was less transient, partly due to the loyalty of its residents, but also because some two hundred men were presently employed in the mines. Contrary to *History of North Idaho*'s assertion that "Wallace threw the weight of its influence and votes in favor of Murray," the November 29, 1897, *Spokane Chronicle* stated that it was the angry union miners' support of Murray that left Wallace a mere 24 votes short of a two-thirds majority. It had received 1,232 votes, while Murray received 652.

By law, the county was precluded from revisiting the issue for another six years, but as soon as the time period expired, the 1892 process was repeated. This time Kellogg was also a serious contender, but Wallace again was the front-runner. Because Osburn's population had decreased, it did not pose much of a threat, nor did Murray, as the general consensus held the county seat should be on the South Side. When the votes were cast on November 8, 1898, of the total 3,335, Murray received 864 and Wallace 2,471. A Board of Canvassers of Elections was required to certify the votes, which they approved, signed, and filed in the Shoshone County auditor's office on November 18th (see sidebar at right). The board was comprised of M. S. Simmons, county commissioner; H. L. Day, county commissioner; and Barry N. Hillard, county auditor.

Despite the voters' approval, and the duty of the county commissioners to move the records to Wallace in a timely manner, the records and county offices remained at Murray for awhile longer. A transcript of a hearing before the U.S. Industrial Commission on July 28, 1899, (in regard to the recent labor troubles in the Coeur d'Alenes) contained an explanation about the delay in the move. It was simply a matter of practicality. The snow was eight feet deep over the summit, making it impossible to move the safes. Plus, Wallace was not yet ready to receive them, as vaults had not been built, nor was the county jail ready to receive prisoners. In addition, all the county officers were in Murray. By law, the county commissioners were to hold regular meetings at the county seat, or wherever the county records were kept, on the second Mondays of January, April, July, and October of each year. Because the records were still at Murray, upon the advice of the county attorney, the January 9, 1899, commissioners meeting was held there, not at Wallace. It was the final meeting for those commissioners, as their terms expired that same day. Consequently, the incoming commissioners inherited the responsibility for the move, and later held a special session to arrange for the removal of the county records to Wallace, where they have remained ever since.

Certified Election Results Authorizing Removal Of the County Seat from Murray to Wallace

On record in the Shoshone County Recorder's Office is a document, titled "Proceedings of the County Commissioners, Shoshone County, Acting as the Board of Canvassers," filed on November 18, 1898, which shows the certified results of the votes on the county seat issue from the November 8th general election:

The Board convened pursuant to law to canvass the votes cast at a general election held in said County on the 8th day of November A.D. 1898, and the following proceedings were had, to-wit:

The Board then proceeded as a Board of Canvassers of Elections to canvass the vote of a General Election held in Shoshone County, State of Idaho for Congressional, State, Legislative, County and Precinct Officers, and upon the question of the removal of the County Seat to Wallace, in Shoshone County, State of Idaho.

Present Commissioners M. B. Simmons
H. L. Day
Clerk Barry N. Hillard

And upon the Canvass of the returns from the several election Precincts, it was found that the following question "For Removal of the County Seat to Wallace," received the number of votes for and against . . . as follows, to wit:

	Yes	No
Littlefield	**0**	**61**
Murray	**2**	**238**
Eagle	**5**	**20**
Delta	**5**	**68**
Burke	**317**	**45**
Gem	**266**	**28**
Mullan	**221**	**75**
Wallace #1	**359**	**10**
Wallace #2	**388**	**26**
Osburn	**29**	**16**
Kellogg	**178**	**50**
Wardner #1	**163**	**76**
Wardner #2	**285**	**81**
Kingston	**110**	**10**
Elk Prairie	**7**	**7**
St. Maries	**0**	**9**
Oro Fino	**25**	**17**
Lolo	**40**	**14**
Weippe	**23**	**9**
Pierce City	**48**	**4**
Totals	**2471**	**864**

Thoughts and Planning for the Court house in Wallace

On March 18, 1905, and March 25, 1905, two newspaper articles appeared, both in the Idaho Press. Both of these address the cost and paying for a new courthouse.

The following appeared in the March 29, 1905 Idaho Press:

A COUNTY BUILDING

FACTS AND FIGURES WHICH INTEREST THE PUBLIC. JUST HOW BAD WE NEED IT

Absolutely No Security at Present For All the Valuable County Records.

Wallace, Idaho, March 24. To the Editor of The Press:

To, build, or not to build – that is the question for us to consider. Fortunately, it is not a political, but financial question.

That we need a courthouse is axiomatic and needs no discussion. All we have to do is to use our eyes on the building now used for a court house and jail purposes. When we touch the pocketbook it brings forth serious reflections, and well it should.

But sometimes we are penny wise and pound foolish.

The question is now up to us as to whether the county shall be bonded for $76,000 to put up a creditable building, and as some of us who have our offices in the present building may know more of the condition than others, let attention be directed to some of its defects.

Present Danger to Records.

The county auditor and assistants occupy the first floor in the front, and in the rear is the vault that contains valuable records and documents of Shoshone county. Not only do these records contain a copy of the deed of your home, but deeds of all valuable mining properties, some of them worth millions, Stop and think of the tremendous loss if these were destroyed. This vault is between the first and second floors, and in its situation it cannot be secure. Directly over this vault is the six or seven ton safe in the assessor's office. In case of fire it must of necessity crash through the vault. That would be the end of all valuable records.

Go into the sheriffs office. You will find all his records upon shelves unprotected. In the office of the probate judge you will find a safe full of the latest records of that office and the oldest records piled on top. The writer believes this safe would not stand a severe fire. You will also find here all the files unprotected on shelves.

In the treasurer's office also the books are unprotected.

The County Bastille.

Then go into the jail. Well, Hicks is a good follow and all that. I would advise you to plead guilty at once and take your sentence to the penitentiary rather than stay overnight in that hole. If exact condition of the old building should become known the County officials would be liable to indictment for cruelty to animals.

Something About the Cost.

Let us look at the cost. That is what will determine our decision April 8. The commissioners deemed it wise to call for bonding the County for $75,000. Some think we can build for much less; perhaps we can. But don't you know our County board will not spend more than necessary. If all the bonds are not needed they will not be issued. Some think it will take not more than $50,000, but say it will take $65,000. From that *sum we can deduct the selling price of the present County building, conceded to be $20,000. This leaves $45,000, I understand this County building was purchased from the expense fund. If that's to be so the money would have to be returned to that account when the building is sold. In any event it would go toward paying off the County in indebtedness.*

The Interest Figured.

The interest on this $45,000 at 5% would amount to $2250 per year. It would not be fair for the County is now paying $800 a year for the courtroom. This must be deducted from the $22,000, leaving $1400.50. From this we can deduct a saving insurance, and with other sums saved we can safely figure our taxes for interest not to exceed $1500 a year more than at present. To offset that amount we will have a good County building, with perfect security for the records.

Payment of Principal.

Now, about the taxes when the bonds are due. Let's look at the proposition.

This County during the past 10 years has grown from an assessed valuation of $1,700,000 to over 5 million. And you are not so pessimistic as to think it will not continue to increase even in a larger ratio during the next 10 years. But let's be conservative, let's say that in the next 10 years we will have and assessed valuation of $10 million. I understand the bonds are to be paid at the rate of 20% each year. 20% of $45,000 is $9000. That would call for an assessment on 9/10 of the mill on the dollar; or, if your taxable property was $500 you would pay on courthouse bonds each year for the sum of 45 cents or thereabouts.

A Broader View of It.

But should we look only to the financial side of the question? The County needs the building and must have it sometime, and when built it must be paid for. So why not have it now and then the next generation pay a portion of the expense. We must keep the records of this County for all that must follow us. So why not cast upon them a part of the burden of providing a place for their security.

Now, let's pick out the convenient and suitable place and put up a good building – one we will not be ashamed to show our friends and neighbors – one that will last as long as the beautiful hills surround us.

J. H. Boomer

The next article regarding a new Court House for Wallace appeared in the Idaho Press on April 8. 1905:

NOT TO BE HELD UP

GOOD AND CHEAP SITE FOR COURT HOUSE CAN BE HAD SO SAYS AN APPRAISER

The Pressing Necessity for a New County Building and Jail Fully Discussed.

The article below is from the ***Murray Sun****, of which Adam Aulbach, one of the courthouse site appraisers, is editor. He is in a position to know exactly what he is talking about and what he says may be considered absolute assurance that the county can secure a building site for exactly its worth, and no more. In this last issue on April 1, the* ***Sun*** *says:*

"The appraisers appointed by the board of county commissioners to put reasonable values on certain sites in the city of Wallace for new courthouse and jail performed their task on Thursday and Friday of last week and submitted the report to the commissioners in a sealed envelope. The appraisers examine 10 sites that had been suggested in different parts of the city, and they found a number available, some more desirable than others. No combination of less than six lots was considered. Until the commissioners see fit to make the report public we cannot give details but we can state this is in all fairness to

the property owners of Wallace that, with one or two exceptions, there has been a known disposition on their part to hold the County up because it wants a few laws for a new building. The appraisers made a careful review of their labors, and recommended a site to the commissioners, but the commissioners are not bound to accept the recommendation, the appraisers being only an advisory board. At in any rate we can assure the taxpayers that the ground room for a new building can be secured at a minimum figure, not a real estate agent's fancying maximums. In the main, property owners themselves were seen and consulted as to price, and when that was obtained there was little need of further parley. The appraisers are satisfied with their labors, and we believe they went to their homes with a pretty good opinion of the Realty owners of the County seat."

Now it is up to the taxpayers to see whether they want a new courthouse and jail or not. One thing is certain, the county must have a suitable building to do its business in and it must have a jail, free from the danger that is ever a menace to the present one. Russia is censured the world over for the treatment of her prisoners. What may not be said of Shoshone County which incarcerates her unfortunates in a fire trap, a mere wooden shell surrounding steel cages, from which no person can be taken in case of fire. The County, having full knowledge of the danger from fire to the present jail, would be liable for large damages to the relatives of prisoners incinerated there. But the everlasting disgrace upon the people of the county would outweigh all money consideration in case of a Holocaust.

The present facilities of the County building are also inadequate and there is no more room and the records vault is already Chuck – a – block, not room enough for another book. All the offices are more or less cramped and badly ventilated, and expense of maintaining them is nearly double what they would be in a new building.
But this whole matter is up to the taxpayers. They are the ones to vote for or against the project next Saturday, and we trust a full vote will be polled in every precinct, in order that the matter may be settled comprehensively, for elections are very expensive luxuries.

The next article appeared in the Idaho Press on April 22, 1905:

NEW COURT HOUSE

SEVERAL MATTERS OF INTEREST CONCERNING IT.

THE DONATION OF THE SITE

Satisfies Nearly Everybody in the County – Fireproof Building Being Considered.

Had it been known before the board election that the courthouse site would be donated to the County is doubtful if there would have been as many as 50 votes against the bonding proposition. Without entering into a discussion at this time as to whether Wallace did or did not promise a site for the courthouse at the time of her efforts to secure the County seat some years ago, it is nevertheless true that she would have loudly promised it had been necessary in order to get the seat of County government, and the impression quite generally prevailed in other precincts of the County that she did promise. The donation of the site was certainly an act meritorious in every respect, and the board of commissioners very properly lost no time accepting it.

The site itself will probably satisfy many more people than would any other selection. Some few objections are brought against it, but not the number and not so serious ones as would apply to other locations. It is the highest and best building ground on the flat upon which the city is built and has streets on three sides of it. It does not, however, get the amount of sunshine some other locations receive. It was higher-priced ground than other offering, but that objection was swept aside by the gift to the County. All things considered, the courthouse site prob-

The DeLashmutt Building, built in 1890 at the northwest corner of Sixth and Bank streets, served as the Shoshone County Courthouse from 1898 until 1905 when a courthouse was built one block to the east. It was also used as the post office after the first one was destroyed in the 1890 fire. Note the sign partially blocked by the pole that reads "Old Court Rooms" These were boarding rooms for women.
(Courtesy Butch Jacobson)

lem is solved in a most acceptable manner to the people of the county in general.

Plans to be Considered April 25.

Four architects have already inspected the site and are preparing plans for submission to the commissioners. At the special meeting on the 25th of this month, the board will consider any and all plans and as soon thereafter as possible will make selection. 30 days from April 15 will be required for advertising for the sale of the bonds.

Fireproof Building.

As nothing definite as to the courthouse building has been determined upon, there is an impression that an absolutely fireproof building will be erected, with not a particle of wood in its make-up. It is said that the sums saved by the donation of the site, if added to $60,000 or whatever sum is determined upon for a building of ordinary construction, would make it perfectly fireproof, absolutely free from the use of any combustible material. If this be true there was hardly a question that the board will give the matter very serious consideration.

To Hold or Sell.

The views held by a number of businessmen of Wallace are that the County cannot do better than to hold in rent the present County building after the new courthouse is built and occupied. Also, it is argued that while the old building and lots would probably bring $20,000 cash, with some repairs and changes, readying the building for business purposes, would be much more than interest on $20,000, and indeed it would be very likely bring sufficient revenue to pay interest on the full bonded debt for the new courthouse. It is one of the eight choicest business *corners in the city, will always be of good value, and no doubt the board will study well what is best in the matter before making final disposition of it.*

On January 25, 1907, ***The Times*** (Under newspaper titles in Shoshone, Idaho, there is a newspaper called "The Times," which was published in Wallace, Idaho from 1905 to 1910) **wrote:**

COURTHOUSE WILL BE READY MARCH I

CONTRACTOR IS EXPERIENCING DIFFICULTY IN GETTING MEN.

This week will see the completion of all of the county court house with the exception of the marble work on the stairways," says superintendent Dahl, in charge of the construction of the county building, representing Contractor August Ilse of Spokane.

"If the men were here now the entire interior could be completed within two weeks, but the trouble is to get the men. They may not be found for some time yet, as so far they can not be located."

The interior of the new court house presents a beautiful sight and when finished, Shoshone will have, one of the of the most complete and attractive buildings in the state. Two months ago the work on the stairways, through some misunderstanding between the contractor and the architects, Stritesky & Sweatt of Spokane, was delayed and no agreement was reached for several weeks. The men who were doing the work in this time had gotten various other jobs and now the trouble is to locate them. Considerable of the delays which have occurred so often since the building was started nearly two years ago have been through circumstances and happenings unavoidable and unforeseen by the contractor. But it is practically certain that the first of March will see the building ready for occupation.

On June 1, 1907, the following article appeared in *The Times*:

TO SOON OCCUPY NEW COURT HOUSE

In Two Weeks It Is Thought the County Offices Will Be Moved.

TO MOVE THE JAIL FIRST

Work of Moving the County Jail Will begin Just as Soon as Judge Steele Concludes His Few Days of Court Here – New Building Is Furnished in Latest Pattern.

The end of another fortnight will probably see the offices and officials of the new county comfortably housed in the new courthouse. Architect L. R. Stritesky of Spokane, who has charge of the final work on the building was in the city yesterday in connection with the changes to be made.

"The business of moving the cells of the county jail into their new quarters will begin, immediately following the closure of the special term of court, to be held next week In the new courthouse," remarked M. Stritesky last evening. "The removal of the jail must be done first, as the noise incident to the putting of the cells into place would so disturb the officials that they would be unable to accomplish anything. As soon as the jail lodgings are in, which will probably by the first week after next, the work of removing the books and equipment of the offices will go right along and be over as soon as possible."

The special term of court, which will be held by Judge Steele of Moscow, beginning Monday, will last for only a few days, so that the transfer of the cells can begin possibly by Thursday or Friday. The officials are looking forward with pleasure to the time when they will be installed in their new quarters and have the use all the conveniences which go with the new building. The rooms are all arranged-so as to be furnished with plenty of light. The furniture is the latest in pattern, being of steel color so as to imitate wood, and will last for many years, being practically indestructible. The interior with its marbled floors and stairways and columns presents a beautiful sight, and the county employees will have reason to be proud of their home, instead of having to apologize every time a visitor comes and say that we have a better building up the street."

The matter of the disposal of the old county building is one that will have to be taken care of by the county commissioners.

Also, on June 1, 1907, another article appeared in *The Times* describing the arrangement of the new court house:

ARRANGEMENT OF THE NEW COURT HOUSE

Where the Various Office and Courtrooms Will Be Located.

JAIL IS IN THE BASEMENT

District Court Room Is On The Third Floor and is Conveniently Arranged With Judges Chambers, Jury Room and Clerk's Office. Fourth Floor for Store Room.

In points of elegance of appearance and simple convenience the new County building which is in process of fitting up for the occupation by the County officials, is one which will surely make the persons who are compelled to live in it glad of their surroundings. The entrance and exits are nicely arranged and the situations of the various offices are convenient for the use to which they are intended to be put.

There are three entrances to the building. The main entrance with its huge doors and beautiful portico and stairway opens from Bank Street. The second entrance and the one which will be used by the patrons of certain offices quite extensively opens off seventh Street and though large and convenient, is not so imposing as the main entrance. The third entrance and the one which will be used by certain of the enforced inhabitants of the court house basement, is located in the rear stairway. This will be used for taking prisoners from the jail to the Superior Court room by a private way.

There are four floors above ground and the basement. The different stories are all nearly fitted up, the halls are large and roomy and present a pretty sight with their marble floors and stairways and columns.

Jail in the basement

The basement is were the County jail will be situated. The room in which the cells will be placed is in the North East corner with windows from two sides so that there would be plenty of light. Immediately behind the general room is the jailer's office where visitors will be received and books kept. In the rear of the office is the kitchen where the meals for the prisoners can be prepared. At the rear of the kitchen and in the south east corner are two cells for women prisoners. The office of the surveyor will be a large room at the corner of bank and seventh streets containing a large vault for valuable papers, etc. The boiler and fuel room will occupy roomy quarters at the rear of this office. In the rear center of the basement is a large storage vault.

On the main floor is the County Treasurer's office occupying the room at the Bank–Seventh streets corner. The assessor's office is in the rear of the Treasurer's office. The County commissioners have a convenient room in the North East corner while the County auditor and recorder will have the room in the rear of the commissioners room. A private office is fitted for the recorder opening off the main room. There are large files in each of the offices for books and papers.
The prosecuting attorney will have his office on the second floor at the Bank – Seventh street corner while the County school superintendent will occupy the room at the office at front corner.

The sheriffs office is located at the rear of the

attorney's office and the probate judge's office and court room in the rear of the superintendent's office. Private quarters are arranged for the probate judge opening off the court room.

District Court room.

The District Court room where the sessions of the District Court will be held is on the third floor in the North East corner and is fitted up in the most modern manner. In the rear of the courtroom is the jury room where the jurymen can retire when they desire to deliberate or when it is necessary that they withdraw for a few minutes. This room is at the Hotel-Seventh streets corner and here the witnesses can remain until they are called to testify. The chambers for the District Court occupy a position in the rear of the witness room and the judges private room will be in the rear of those chambers. The clerk of the District Court will have an office in the center of the rear of the floor. The fourth story while large and roomy is not so well lighted and will be used at present as a storage room and attic with the exception of one room in the rear which is for the purpose of keeping jurors, should they find it necessary to deliberate over a question overnight.

Large toilets for both sexes are placed on each of the first three floors and the basement floor.

Transfer of prisoners

The question of where the County prisoners will be kept during the several days that will be required for the removal of the cells to the new building and they're putting in place has not yet been definitely settled. The work of removing the cells and setting them up will be taken care of by The Coeur d'Alene Ironworks, as is necessary to use the services of some boilermakers. The County building which was designed by Stritesky and Sweat of Spokane is now in the charge directly of architect Stritesky, the firm having dissolved partnership. Mr. Stritesky will have direct charge of finishing up the work.

The Shoshone County Courthouse at Wallace. *(Courtesy Butch Jacobson)*

Levi and May Hutton in Wallace
and Jamie and Barbara Baker's Wallace connection

L. W. "Al" Hutton
1860 – 1928

As a partner with the original discoverers of the "Mighty Hercules," one of North Idaho's richest and most famous silver mines, L. W. Hutton, was to become a legend – not for his wealth, but for how he used it. Throughout much of Spokane's history, Hutton left a mark of goodwill towards his community that remains unsurpassed. He was truly an honorable man.

Levi W. Hutton, the son of Levi and Nancy (Holsinger) Hutton, began his life at Fairfield, Iowa, on October 22, 1860. Less than a month after his birth, on November 11, 1860, Levi's father died. Almost seven years later on October 7, 1867, his mother died, leaving him and his six siblings to be distributed among relatives, never again to be reunited as a family. While living with his uncle in Iowa, his burden of chores was noticeably greater than that of his cousins. While his cousins attended school, Levi's farm chores limited his education but provided him with an overwhelming appreciation for the hardships of others.

The former home of Levi and May Hutton at 229 Pine in Wallace. This residence in now owned by Jamie and Barbara Baker, who are standing on the front porch. During a trip to Wallace in 1903, President Roosevelt was a house guest with the Huttons at this residence. *(Bamonte photo)*

At the age of 12, Levi set out for the gold fields in the Black Hills of South Dakota. Finding that adventure too demanding for his young age, he soon returned to Iowa. Shortly after, as a teenager, he left Iowa to live with relatives in Salem, Oregon. In 1881, at the age of 21, he drove a four-horse team from Portland, Oregon, to Spokane Falls. From Spokane Falls he moved to Lake Pend Oreille, where he spent almost a year working as a steamboat fireman, loading cord wood onto the steamer and tending the firebox for the steam boiler. With his freshly gained knowledge of steam-engine operations, he moved back to Spokane Falls and was hired by the Northern Pacific Railroad as a locomotive fireman.

Upon completion of the Northern Pacific's transcontinental line in 1883, Hutton was transferred to its Missoula Division and promoted to the position of engineer. In conjunction with, and as a result of, the completion of the railroad, a placer gold strike in the Coeur d'Alene Mountains was announced, and a rush to that area began. At that time, L. W. was transferred to the Wallace Division, specifically to Wardner Junction (now Kellogg).

Hutton's public legacy began when, in 1887, he met his life partner, May Arkwright. May had arrived in the Coeur d'Alene Mining District at the outset of the gold rush in 1883. She subsequently moved to Wardner Junction around 1886. May was reputed to be the best cook in the Coeur d'Alenes, and Hutton became a regular mealtime diner. Levi apparently did not care for his given name and preferred, as was customary of the era, to be referred to by his initials,

The former Hutton home on E. 2206 17th Avenue in Spokane. *(Bamonte photo)*

L. W., or "Al," as he was affectionately called by May and his close friends. The explanation for his nickname "Al" may have been the sound of "L."

By 1895, L. W. and May had saved enough money ($880) to purchase a 3/32 interest in the Hercules, a long-shot mine that had been discovered by Harry Day and Harper six years earlier. Besides the initial capital the Huttons invested, they also made a commitment to perform manual labor on the claim, with no guarantee of any return. On June 2, 1901, fellow investor August Paulsen blasted into a solid vein of the richest silver ore that had been discovered up to that time in the Coeur d'Alene Mining District.

The riches from the Mighty Hercules afforded the Huttons means to accomplish many things, and together they chose to use their fortune to help others. Perhaps L. W.'s early life as an orphan left him with great understanding and appreciation for the value of integrity and kind consideration, leading him to his greatest philanthropic achievement, the Hutton Settlement, established in 1919. Although May died before L. W. announced the Hutton Settlement as a repository for their combined fortunes, they both held a mutually blessed goal of helping children in need.

On November 3, 1928, at the age of 68, L. W. Hutton died as a result of complications from diabetes. In his will he left the bulk of his estate to the Hutton Settlement. His funeral at the downtown Masonic Hall, with the Knights Templar officiating, was attended by almost 2,000 people. His body was laid to rest beside his wife's on a high point of land overlooking the Spokane River at Fairmount Memorial Park, a spot they had chosen together years earlier.

May Arkwright Hutton, 1860 – 1915

Mary "May" Arkwright Hutton was born on July 26, 1860, at Washingtonville, Ohio, a small coal mining town. At the time of May's birth, her father, Isaac Arkwright, was married to Catherine Arkwright. Together they already had four children. Outside of his marriage, Isaac fathered May with a woman who either died or abandoned her young daughter. At the age of nine, her father removed her from school and sent her to live with her blind grandfather, where she served as his housekeeper, guide, and companion. May's grandfather was well-connected politically, and she often escorted him to numerous political meetings around town. As a result of her grandfather's influence, May became indoctrinated at an early age to the plight of the less fortunate.

In 1883, May joined a small contingent of coal mining families from the Washingtonville area, who decided to relocate 1,800 miles across the United States to the first gold rush camp in the Coeur d'Alenes, a primitive setting called Eagle City. Here, May immediately gained employment as a cook. Within a short time the bulk of placer gold was exhausted and the rush moved approximately five miles east to a new and richer gold-bearing area named Murrayville (soon shortened to Murray). In 1884, when the population shifted from Eagle City to Murrayville, May immediately went to work for Jim Wardner, a local entrepreneur she had met during her 1883 Northern Pacific railroad trip to Idaho. Wardner had procured a saloon on the main street of Murray and set up a food counter in a back room, which May operated. During this time, she lived in a lean-to attached to the rear of the building.

By 1886, Jim Wardner had become a major player in the Bunker Hill and Sullivan mining venture, establishing Wardner Junction (later renamed Kel-

May Hutton, far left, with some of her friends in front of the Hercules Mine portal, circa 1905. *(Courtesy the Hutton Settlement)*

logg), the site of a new railroad station. Wardner influenced May's move to that location, where she opened her own boarding house. During the spring of 1887, May met Levi W. Hutton, a locomotive engineer who regularly ate at her boarding house. They married on Thanksgiving Day, November 25, 1887. A few years later, they became suddenly wealthy from their involvement in the richest silver strike of the times in the Coeur d'Alenes. Their philanthropic spirits quickly surfaced, and they became two of the greatest champions for the poor, oppressed, and underprivileged in the Inland Northwest.

During the early years of their marriage, the Huttons lived in Wallace, Idaho, amidst a violent upheaval between the unions and mine owners. The Huttons were pro-union and champions of the underdog in these struggles. When the unrest turned violent in 1899, masked strikers commandeered L. W.'s train at gunpoint to haul the dynamite and men to blow up the concentrator mill of the non-union Bunker Hill and Sullivan Mining Company.

Following the explosion, Idaho Governor Frank Steunenberg sent an urgent request to President McKinley to immediately dispatch 500 regular troops to the Coeur d'Alenes. Martial law was declared, and military troops quickly and efficiently rounded up and arrested about 700 rioters, who were forced to build a stockade (known as the bull pen).

L. W. was among those arrested and confined in the stockade. By relentlessly badgering the guards and the governor's on-site representative, May was able to secure her husband's release. This was to be a preview of May's persuasion and tenacity.

Eugene R. Day's Hercules teaming outfit at Burke in 1911. Eugene's brother Harry and partner Fred Harper discovered the Hercules in 1889. Harper sold his one-half interest before the true value of the mine was revealed in 1901 by August Paulsen. *(Courtesy Butch Jacobson)*

May helped Idaho women gain the right to vote in 1896 and, in 1904, ran as a Democrat for the Idaho State Legislature. When the Huttons moved to Spokane in 1907, May became heavily involved in Washington's suffrage movement. She wrote and lectured tirelessly, organizing and campaigning towards the 1910 state election that gave Washington women the right to vote. Upon reaching that milestone, May is claimed to have been the first Spokane woman to register to vote. She was also one of the first two women to serve on a Spokane County jury. In 1912, she attended both the state and national Democratic conventions.

In addition to being one of the most politically active and famous women in the Northwest in the early 1900s, May Hutton was also one of the kindest and most ambitious. The Florence Crittenton Home for unwed mothers, the Spokane Children's Home, and other similar charitable endeavors were recipients of May's time and wealth.

Following a long bout with Bright's disease (a kidney condition), on October 6, 1915, at the age of 55, May died in her home at 2206 East 17th Avenue. Al, in loving remembrance, designed her tombstone, half rough and half smooth, to signify a life half-finished.

The Hutton Settlement
9907 East Wellesley Avenue
Spokane, Washington

When L. W. Hutton's partners discovered one of the richest silver ore deposits of the times, it was as if that discovery was deliberately intended for the kindest of people. When Nature's vault, deep inside a mountain, was opened for the Huttons, that wealth would produce a volume of good works, the pinnacle of which would be the Hutton Settlement, a home for children who were either orphaned or from troubled or abusive situations.

There were also a number of other buildings and structures in Spokane that came from the wealth from the Coeur d'Alenes. Among them were Amasa Campbell (the Campbell house), The Finch house, Patsy Clarks (the Clark Mansion), and many investors for the Davenport Hotel were involved. However, no person whose wealth came from the mines in the Coeur d'Alenes, has ever matched or even came close to the kindness shown to the community by Levi Hutton with his everlasting and perpetually sustainable gift to children.

On August 28, 1917, in a front page headline, the Spokane Daily Chronicle announced L. W.'s long-held ambition to construct the Hutton Settlement. His subsequent purchase of 112 acres from local railroad magnate Daniel C. Corbin was the first step toward a dream of helping children in need of a home. His goal was to create a home-like atmosphere where brothers and sisters would not have to be separated, as he had been from his siblings.

With Harold Whitehouse as his architect, the Settlement was constructed to accommodate up to 80 children, who would be able to enjoy a normal lifestyle in a loving environment. He spared no expense in the construction of the large late-English-style brick homes, arranged in a campus-type setting that also incorporated an operating farm. He also established a trust for the Settlement that would ensure its existence as a children's home in perpetuity. By the time the Hutton Settlement was completed and occupied in 1919, he had purchased additional land, bringing the total to 320 acres.

No more fitting tribute could be paid to the Huttons than a simple statement made by one of the residents several years ago. A young girl was asked, "What is the most significant or nicest thing you know about the Huttons?" Without hesitation and with total sincerity, the young girl replied, "The Huttons made a home for us."

A considerable honor to L. W. occurred in 1920, when he was invited to the National Shriners Convention in Portland, Oregon. As the original founding member of the Shrine Blue Lodge in Wallace, Idaho, and an active member of the El Katif Shrine in Spokane, L. W. was recognized at this conven-

The famous Hutton Settlement, circa 1921. In 1918, Levi Hutton commissioned Spokane architect Harold Whitehouse to design the complex. Hutton spared no expense in the construction of the settlement, which includes four brick-faced Tudor Revival cottages, each designed to house up to 30 children, on a college campus type of setting. The campus is surrounded by farmland, which was used in the early years to grow gardens and raise animals to help provide food for the settlement. Upon Hutton's death, virtually his entire estate was placed in a perpetual trust endowed to the settlement. The Hutton Settlement continues to provide a home for deserving children in need of a safe and healthy environment. It is a living tribute to what wealth can accomplish in the hands of compassionate and caring people. *(Courtesy the Hutton Settlement)*

tion by National Imperial Potentate Frieland Kendrick. Referring to Mr. Hutton, he addressed the crowd: "If one man could build and do for children what Mr. Hutton has accomplished, what could 500,000 Shriners do?" That convention and comment spawned the concept for the Shriners' charitable arm, which is the Shriners Hospitals for Children. Its first hospital opened in Shreveport, Louisiana, in 1922. Today, the Shriners Hospitals include a network of twenty-two hospitals in the United States, Mexico and Canada.

This monument is an expression of the indebted respect and appreciation to L.W. and May Hutton for their considerable and unending contribution to the youth of the Inland Northwest. Their unseen hands continue to guide countless children into productive adulthood. This monument is dedicated to one of Spokane's most beloved and honorable couples.

The Huttons' Lasting Spokane Connection Courtesy Al and May Hutton From Wallace, Idaho

The Hutton Building, on the east side of Washington between First and Sprague, in 1920. It was designed by Dow and Hubbell and completed in 1907 as a four-story building. It initially became known as the Chamber of Commerce Building as the chamber leased the second story for five years. The Huttons moved to Spokane the year the Hutton Building was built. They lived in a comfortable apartment on the fourth floor until 1914, when they moved to a home they had built on 17th Avenue. Three additional floors were added in 1910. The building north of the Hutton Building was the Lindelle Block, built shortly after the great fire of August 4, 1889, and demolished in 1963. Just beyond it (the tallest building at the left) was the former Spokane Club, by this time known as the Chamber of Commerce Building. *(Northwest Museum of Arts and Culture, L87-1.17866-20)*

Staff from the Fairmount Memorial Association, Orville Clouse, Duane Broyles and granddaughter Sierra Broyles, Steve Pratt, Dave Clark, Tom Stokes, Kevin Smith, Carl Ellis, Dave Peters, Alex Crosen, and Diane Perry surround the first three-column monument created in honor of Levi and May Arkwright Hutton, who were partners in philanthropy. Their legacy lives on in the Hutton Settlement, dedicated to helping children for nearly a century. Located in Spokane's Fairmount Memorial Park, the memorial was dedicated on April 3, 2008. *(Bamonte photo)*

Levi Hutton and children at the entrance to the Hutton Settlement Administration Building, circa 1921. *(Courtesy the Hutton Settlement)*

This photo was taken on Father's Day the year after Levi Hutton's death in 1928. It was held at the Huttons' graves at Fairmount Memorial Park. The Settlement's Boy Scout troop and a Camp Fire girl bugler led a grave-side service in his honor. In 2008, a monument was erected at this site by the Fairmount Memorial Association to commemorate the untold contributions made by the Huttons during their lifetimes, which continues through the ongoing service of the Hutton Settlement. *(Courtesy the Hutton Settlement)*

Wallace's First Hospital

Wallace's first hospital was called the Frances Holland Memorial Hospital. The cornerstone for this facility was laid on May 20, 1890, and it was ready for occupancy Oct. 1, 1890. The Holland Hospital was built by Holy Trinity Episcopal Church and operated by the church for about one year. It was then leased by a Dr. Sims but was sold to another doctor at a later date. It was damaged by fire about 1913, but rebuilt and called the Wallace Hospital (see photo on following page). At the time of this picture, the cost to be a patient in the hospital was $1.50 per day. Miners had $1 per month deducted from their pay for hospital services when needed for medical care at the Providence Hospital also. *(Courtesy Butch Jacobson)*

The Following article is from the *Wallace Free Press* of May 17, 1890:

James Holbrook with his wife, (Archdeacon Gunn's sister) and three children have arrived in Wallace for the purpose of taking charge of the hospital about to be erected by Archdeacon Gunn at the west end Cedar street, at the foot of the mountain which stands like a wall in that part of the town. It can be made a beautiful spot by clearing away the forest and leveling off the tract, a little over an acre of ground. Placer creek, with its clear, waters, separate the site from town. The object of the hospital is to take care of the sick and wounded of the mines and railways about Wallace. The property will belong to the Episcopal church, and the work will be under the sole charge of Archdeacon Gunn, but there will be no difference shown between persons of creed, color or nationality. By the payment of $1 per month a person can secure a hospital care, medicine and medical attendance. The hospital will be known as "Holland Memorial Hospital," as it is through the generosity of Mrs. Frances Holland of New York that Archdeacon Gunn will be able to build it, and he very properly names it in memory of her late husband. Mrs. Holland, we are informed by Mr. Gunn, has been for several years interested in his work in Minnesota and had helped him build seven churches and the Wilder Farm College.

The newly named Wallace Hospital, formerly the Holland Hospital, which was rebuilt shortly after a fire damaged it. Doctors Paul Ellis and Ernest Gnaedinger were the final owners of the Wallace Hospital, which was closed down in 1963. Gnaedinger maintained his medical practice in the building for two years after it was closed. It was torn down in 1971. *(Courtesy Butch Jacobson)*

The Wallace Hospital (above) with a new brick front added on, circa 1933. This building was destroyed circa 1971. *(Courtesy Butch Jacobson)*

The Providence Hospital in Wallace, 1906. The mines and miners had been paying into a fund for a much-needed hospital when they enlisted the aid of the Sisters of Charity, who built Providence in 1892.

On February 8, 1921, it became the birthplace of movie actress Lana Turner, Julia Jean Turner. In 1927, her family moved to San Francisco. At age 15, Lana was approached while sipping a coke at the Top Hat Cafe by W. R. Wilkerson, publisher of the Hollywood Reporter, who had noticed her beauty. With his help and connections, she was on the way to stardom as one of the greatest natural beauties in movie-screen history. She became known as MGM's "Sweater Girl" and was one of its leading ladies. *(Courtesy Butch Jacobson)*

Lower left photo is Lana in Wallace at age five. *(Public domain)*

Lana Turner, born Julia Jean Turner, in Wallace Idaho on February 8, 1921, was a film and television actress. At the age of 16 she was signed to a personal contract by Warner Bros. director Mervyn LeRoy, who took her with him when he moved to Metro-Goldwyn-Mayer in 1938. During the early 1940s, Lana established herself as a leading actress, and was nominated for an Academy Award for Best Actress. Lana starred with many leading actors, here she is seen with Clark Gable. In 1958, her daughter, Cheryl Crane, stabbed Turner's lover Johnny Stompanato to death in their Beverly Hills home; a coroner's inquest concluded that Crane had acted in self-defense. In 1982, she accepted a much publicized and lucrative recurring guest role in the television series Falcon Crest, affording the series the highest rating it ever achieved. Turner made her final film appearance in 1985, and died from throat cancer in 1995, aged 74.

The City Dye Works at 505 Bank Street, Wallace, in 1925. This dry cleaning business was owned by Lana Turner's father, Virgil Turner (far left). When his business failed, Virgil went to work in the mines. Virgil was a miner at the Hercules Mine about the time Lana was born. The family lived in Burke about a hundred yards above the Burke school, before they moved to Wallace. They lived on the upper floor at 217 Bank Street in Wallace. Virgil later moved the family to California. On the evening of December 14, 1930, Virgil's body was found slumped against a wall at Mariposa and Minnesota Streets in San Francisco. Earlier that evening, he hit a winning streak at a crap game held in the basement of the San Francisco Chronicle Building. He had stuffed his winnings into his left sock. When his body was found, his head was bashed in and his left foot was bare. *(Courtesy Wallace District Mining Museum & MGM)*

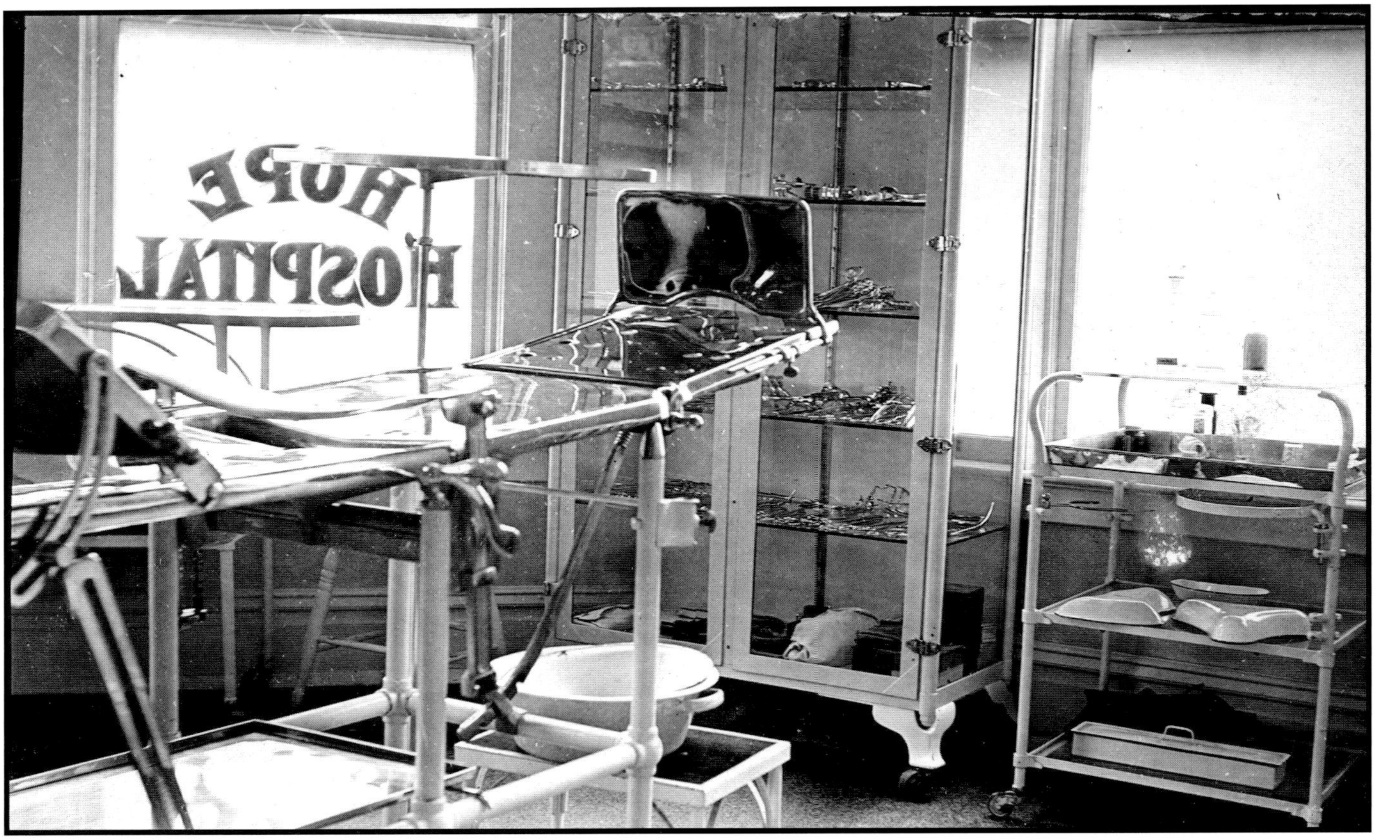

The operating room in Wallace's Hope Hospital, located on the second floor. The Hope Hospital was built in 1905 and located at 500 Cedar Street. The building was acquired in 1916 by the Day Brothers, and used to house their mining operation upstairs. The west end of the building was a grocery store. In the 1920s, Morrows Dry Goods occupied the entire first floor. The building became the Hope Hotel in 1992. This appears to be an operating table surrounded with many of the tools, drainage systems and other equipment used to perform operations and other procedures for any in the area who may need such. *(Courtesy Anderson collection)*

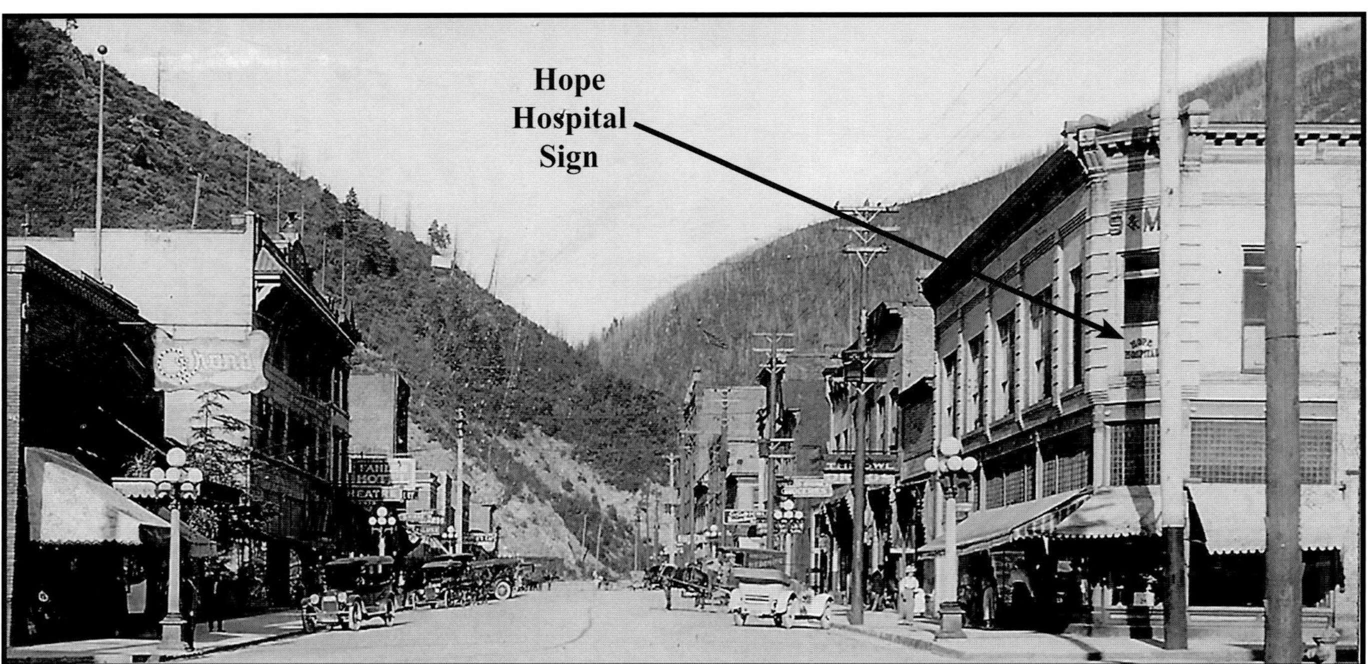

Cedar Street Looking East. The first building on the right is the Hope Hospital, with the arrow pointing to a sign that says "Hope Hospital." It is now the Brooks Hotel building. *(Courtesy Butch Jacobson)*

Mayor Charles Mowery and family in Wallace during 4th of July parade, circa 1915. *(Courtesy Dr. Don Harvey husband of Betty Mowery, daughter of Charles and Doris Mowery)*

Class photo taken in Wallace in 1890. *(Courtesy Dr. Don Harvey husband of Betty Mowery, daughter of Charles and Doris Mowery)*

Doris Mowery with a group of young girls in Wallace. Unknown which girl was Doris, but a good photo of early times in Wallace, circa 1915 *(All photos this page courtesy Dr. Don Harvey husband of Betty Mowery, daughter of Charles and Doris Mowery)*

Dr. Major Herbert Mowery

Dr. Charles Mowery

Doctors Charles and Herbert Mowery's office building at Sixth and Cedar Street, circa 1920. *(Courtesy Butch Jacobson)*

Dr. Charles Mowery in medical school, second row, third from right. *(Courtesy Don Harvey husband of Betty Mowrey, daughter of Charles and Doris Mowery)*

Theodore Roosevelt
The Twenty-Sixth President
Of the United States Visits Wallace

On May 26, 1903, President of the United States, Teddy Roosevelt visited Wallace, Idaho. In this photo he is shown leaving the Great Northern Depot in his carriage. When the president visited Wallace, he spent approximately two hours in the city. The population at that time was approximately 1000 people in the city itself. However, the crowd that attended was 10,000. The day he was in Wallace special trains brought crowds of people to attend this historic event. This was the only time a president of the United States ever visited Wallace. At the time, the main areas that surrounded Wallace were Burke, Gem, Murray, Mullan, Osburn, Wardner and Kellogg. *(Courtesy Butch Jacobson)*

On April 1, 1903, president Theodore Roosevelt began a nine week journey traveling across the United States on a 14,000 mile trip to visit the western part of the country. During that trip he visited 25 different states and gave a total of 263 speeches.

At 9:05 a.m. his special train pulled out of the Pennsylvania Station in New York. As it did, the president stood on the platform of his car tipping his hat and smiling in response to a large group of admirers. Also, to see him off were the members of the legislature and city officials. The train he left in was special – specifically put together for the president. It was described as handsomely equipped, and one of the finest ever run out of Washington by the Pennsylvania Road. It was operated by a crew of

President Teddy Roosevelt's train as it passes through the Rockies in 1903, on his western tour.
(Library of Congress)

handpicked men, in addition, those officially designated as members of the president's party. Three secret service men and two post office inspectors accompanied the president as personal bodyguards.

Amid cheers from thousands of Idaho's loyal citizens the train bearing President Roosevelt pulled into Wallace on Tuesday-morning, May 26, at 7:30 a.m.

The following is part of an exact quote that appeared in the *Idaho State Tribune*, on May 26, 1903. The Tribune was a newspaper published in Wallace, Idaho by the Miners Union, which was in business from 1894 to 1905:

> *Never before in the history of the state has a man been honored as was the president when met at Wallace by the enthusiastic demonstrations of 10,000 rugged souls of this great western country, and 'twas with deep sense of pride that he numbered himself as one of us on that day. He is the only president having visited the Coeur d'Alene mining district, and being the most popular ruler in the world, it is not strange that he should be tendered the splendid reception and greeted with those great demonstrations that made Roosevelt Day in Wallace an event to be remembered through ages to come. Many remarked that it had been years since they had seen such a high spirit of patriotism as was displayed on that memorable occasion. As the President stepped on the platform of his car Senator Heyburn introduced him to the crowd, which responded with loud cheers. The school children laid a path of flowers from his car to the carriages which carried him and his party up town.*
>
> *In the carriage with the president were Mayor Connor, Senator Heyburn and Secretary Loeb. The carriage containing the president and the party was guarded by secret service men, mounted police and 10 Spanish war veterans.*
>
> *Not withstanding the rainy weather that prevailed throughout the forenoon, the city made a beautiful appearance. The streets were hung with thousands of flags of every size and almost every building in the city was lavishly dressed in stars and stripes.*
>
> *The procession which paraded the streets was a sight to inspire the most enthusiastic patriotism. Men from every walk in life dropped into line, forming a procession over a mile long. National airs rendered by four bands, and the continual cheers of the crowd, filled the air with a grand tribute of welcome to the first man of the nation. Every part of the program was successfully carried through. Commendation is due Marshal of the Day Rossi and his able staff for their splendid work.*
>
> *When the presidential party reached Senator Heyburn's home they called upon the Senator and his mother, Mrs. Sarah G. Heyburn, assisted by madames F. F, Johnson, W. W. Woods, Warren Truitt, G. H. Ailshie, B. F. O'Neil, E. H. Moffitt, C. W. Beale, G. W. Kester and Miss Maud Hammell. The husbands of the married ladies were present, also Mayor Connor of Wallace and Senator Ankney of Washington.*

After the reception the president was driven to the grandstand at the park, where be was introduced and presented with a metallic souvenir by Senator Heyburn. The souvenir is pyramid in shape and is appropriately inscribed on the front side with these words: "Presented to Theodore Roosevelt, President of the United States, by the Citizens of the Coeur d'Alene Mining District, at Wallace, Idaho, May 26, 1903." On the reverse side are the words:, "The Gold, Silver, Lead and Copper in this Souvenir were Extracted from the Mines of the Coeur d'Alenes."

Senator Heyburn: "Fellow citizens; I have the distinguished honor of introducing to you this morning the President of the United States. **(Great applause)**. *And, fellow citizens, in conformity with your wishes, I take this opportunity, before the President addresses you, of carrying out your wishes in presenting the beautiful souvenir which you have prepared for his reception. Mr. Roosevelt, in this land where we produce gold, silver, lead and copper, my fellow citizens have thought it appropriate to present you with a testimonial of their high regard in the shape of a little metal monument made entirely from the metals taken from our own mines by our own skilled artisans, in memory of this great day for the Coeur d'Alenes. Please accept it, – also an individual token from one of our gold producing mines, properly inscribed, "Fellow citizens, the president."*

Mr.-Roosevelt: "I feel a little like a'school boy at Christmas time—I want to look at my presents first." ***(Laughter)***

"Senator, and you my fellow Americans, men and women of Idaho. It is a great pleasure to me to be within the bounds of your state, and in greeting all of you this morning I wish to say a special word of greeting to two bodies of your citizens. First of all, to the men of the Grand Army, ***the men who in the Civil War from '61 to '65 proved their bravery and courage.*** *I greet you, not merely because of what you did there, not merely because you left us a reunited country, a country in which wherever the president goes, from the Atlantic to the Pacific, from the Gulf to the Canadian boundary, he may feel at home, but because you left us by your deeds, imperishable memories as to how we should conduct ourselves now, and most vital of all you left us the memory of brotherhood and what brotherhood means; brotherhood, my comrades, with the gallant men against whom you fought, the*

Teddy Roosevelt's courage and leadership during the Spanish American War is widely recognized. Perhaps less remembered is the attempted assassination by a Milwaukee saloon keeper on October 14, 1912.

While campaigning for the presidency under the 'Bull Moose Party,' Roosevelt was shot in the chest. The wound was not fatal, but the surgical removal of the bullet was deemed too risky. Despite his wound, Roosevelt continued his campaign and delivered his scheduled speech. The bullet remained lodged in his chest for the remainder of his life. *(Public domain)* *gallant men in gray who, and whose sons, are now as straight Americans as any of the rest of us.* ***(Applause)*** *And whenever I speak to bodies of Union veterans I speak to men who I know will be first to respond in generous tribute to the men of the South—the men against whom they fought. (Applause.)*

"Then a word of greeting to my own comrades, the men of the Spanish-American war. In that war we had one difficulty from which you of the big war were wholly free; in our case there was not enough war to go around. ***(Laughter)*** *At the same time those who hiked in the Philippines had a very good experience of it.*

"I also want to say a word of greeting to the representatives of the University who are here. In traveling through the great states, these states great in the present, and infinitely greater because of the assured future that looms great before us, nothing has pleased me and impressed me more than the way in which, amidst all the life in exploiting the material resources of the present, such attention is paid to the generation that is to grow up in the future. Everywhere I have been greeted by the school children, and I think you all know that I am particularly glad to see them, and want to compliment Idaho on the quantity and quality of them. ***(Laughter)***

"I have but one word to say to the children: I believe in play and I believe in work. Play hard while you play and, when you work don't play at all, and, that is fairly good advice to the elder folks as well.

"As I said, it has been a .great pleasure to meet the evidences of the care paid to the care and welfare of the next generation through the schools, We must have good material as a foundation. Unless we can have good citizens, good hard workers, men who can get the utmost out of the resources in mines, in lumber, in agriculture, in grazing – in every branch of active work; unless we have that capacity the nation cannot go on, but the capacity is, not by itself enough; on it we must build the superstructure of the broader and higher life; there must be a chance for mental and spiritual growth in addition to caring for the things of the body, otherwise we cannot develop our life and as it should be developed; and I congratulate you upon the care you have given both to elementary education and the higher education of the state; and to the students of the University I have just this to say: Much has been given to you, and from you rightly we will expect much in return. You have had special advantages in the way of education – in the way of training – and it behooves you to show that they have not been wasted; that the state has been justified in bestowing them upon you. You can make no return to your Alma Mater, to your college, save in one way; you can make return for the education and training you have received in the College by that kind of service you render the state.

And The Republic as you grow up. That is the only way in which you can make a return, and we have a right to expect of you the highest type of service that can be rendered by American citizens.

"Now my friends, I come into a state with great natural advantages requiring hard and intelligent work in order that they might be developed. In this state you have mines and forests, agricultural regions, pastoral regions in which irrigation will prove of great advantage, but in this climate and at this end of the state I shall not speak of irrigation. ***(Laughter)*** *I think there would be a certain inappropriateness in it. You have had these great advantages, they would have counted for less than nothing if you had not the type of manhood among your citizens which has enabled you to make the advantages shown in reality as such. Exactly as in war, is the man behind the gun, more than the gun, so in peace no physical quality of the country will unveil if there is not the right type of man to take advantage of it. The advantages count just as good as weapons. We all know, would've been in the Spanish War that it is a bad business to have*

a black powder musket when yellow fellow has a smokeless power one; but if you take a poor fighter with the best weapon in the world he can be beaten by a good man with the club.

"You need a good weapon and the difference will be decisive. Between two men evenly matched the weapon may turn the scale, but more than the right type of weapon you need the right type of man. Isn't that so? It is the same way in citizenship, and the lesson that every American citizen should learn is that all honorable work is noble, and confers honor upon the one who is doing it. The point is not what the work is but to do that work well. If the man be a miner or ranchman, employer or employee, lawyer, banker or mechanic, it makes no difference whatsoever, if he does what his hands find to do truly and well. If he does not do it well, whatever the work is, then he is not a good citizen; if he does do it, and do it well, then he is a good citizen. There are two qualities, two attributes which do more to harm to the citizenship than any others: the arrogance which looks down on those not so well off as itself, and treats them with brutal selfishness and disregard of their interests grow, and equally base espirit de core and hatred of those who are better off. That spirit is utterly un American utterly alien to the spirit shown by the man who fought for the Republic and who, under the lead of Abraham Lincoln saved the liberty of the Republic. ***(Applause)*** *The men who showed that we were a union in fact as well as in name and that this was a government not of license but of liberty, and the worst foe to American citizenship, to American life, is the man who seeks to cause hatred and distrust between one body of Americans and no matter to whom the appeal is made, whether to inflame section against section, creed against creed,, or class against class. In any event the appeal is unworthy of American citizens.*

"Now let me tell you one story, the only one I shall tell. I have been a Western man myself , and used to live in the Little Missouri country. You know what I call country the branding iron takes the place of the wire fence, and if in the round up a calf is missing and not branded when he is a year old, and has no brand, is called a maverick, and by ranch law anyone who found a maverick would put the brand on him according to the ranch on which he was found. Well, I had a new hand one day and in the course of our ride across a neighboring ranch we came across a maverick, and my man got out his rope and we went through the preparatory to branding her. We had no iron and so we would use this cinch ring to run the brand. I said, 'thistle brand' to the man and he started to put on the brand. I suddenly noticed that he was putting on my brand and I says "Here, my man, what are you doing? And he said, All I always put on the boss brand. Well, I says, you go back to the ranch to get your time. The man who steals for me will steal from me just as quickly." Any man who endeavors to make you do wrong under the plea that it is for your interests will in his turn do a wrong to you if the opportunity arises.

"The only safe principle upon which to act is in accordance with that immutable law of decency and fair dealing: to give each man his chance; to give to one man the same justice that you would give to another yo draw the line not between one class of men and another, not between the rich man and the poor, but between a man who is crooked and the man who is straight, without reference to whether he is rich or poor. That is the only ground upon which you can judge a man. ***(Great applause)***

The procession for President Roosevelt in Wallace consisted of 13 carriages. The following is a list of those who rode in the carriages during the time Roosevelt was in Wallace (Approximately two hours):

Carriage Number 1: The President, Secretary Leob, Senator Heyburn and Mayor Connor. One seat for secret service man on box.

Carriage Number 2: Three secret service men and N. O. Latta.

Carriage Number 3: Assistant Secretary Barnes; Surgeon General Rixey and B. F. O'Neil.
Carriage Number 4: N. P. Webster, J. L. McGrew and Dr. France.

Carriage Number 5: Reserved for men of distinction, as governors, United States senators and members of congress, when such accompany the president. Secretary Moody, Judge Truitt and F. F. Johnson.

Carriage Number 6: H. A. Coleman, H.R. Hazard and Maj. W. W. Woods.

Carriage Number 7: Lindsay Dennison, Mr. Morgan and P. F. Smith.

Carriage Number 8: R. N. Dunn, Harry White and N. Lazarnick.

Carriage Number 9: George B. Lockey, H. A. Strechmeyer and Francis Jenkins.

Carriage Number 10: P. W. Williams, J. R. Gooch and C. W. Beale.

Carriage Number 11: President McLean, Mr. & Mrs. Thompson and Mr. McIntosh.

Carriage Number 12: Mr. Fort, Mr. Clark, Mr. Sweet and Mr. Savage.

Carriage Number 13: Mr. Doherty, Mr. Howes and Mr. Kettenbach.

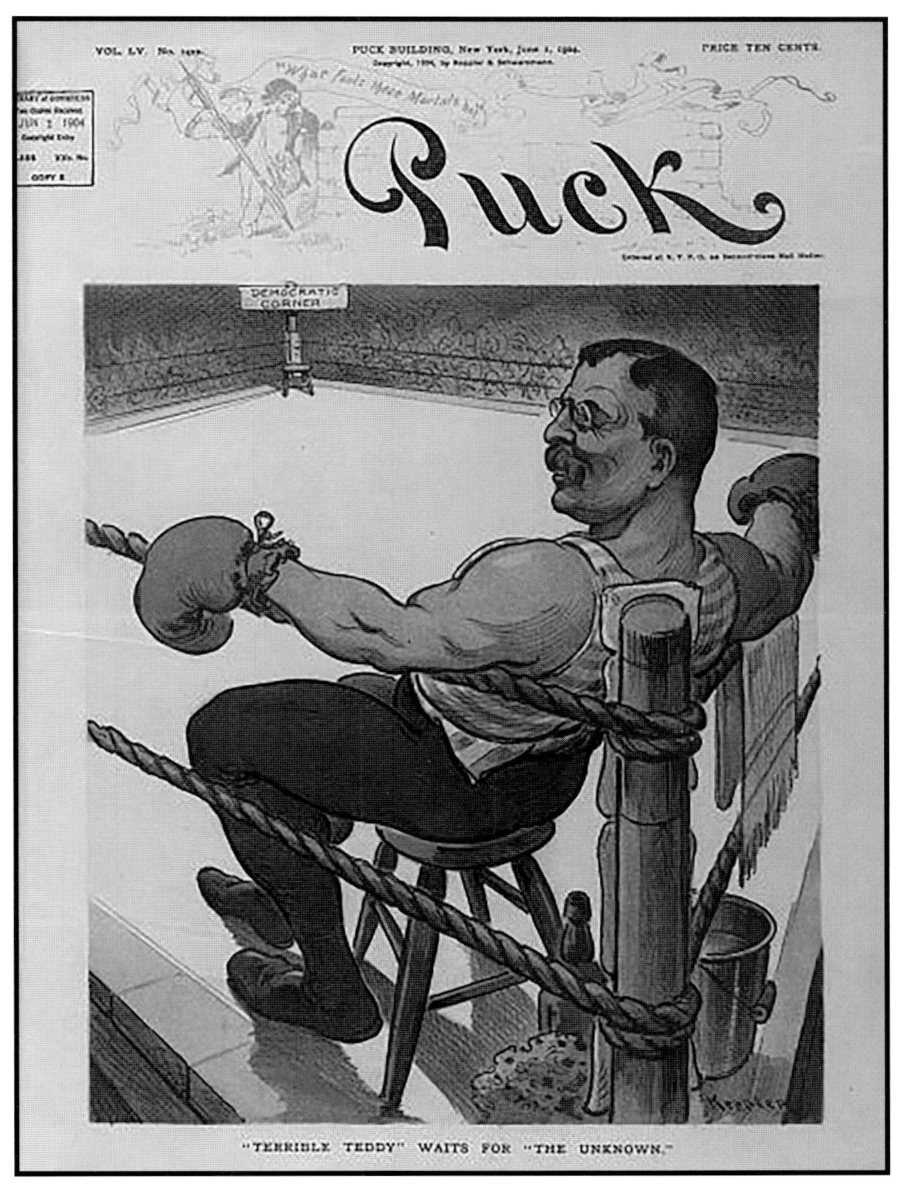

"TERRIBLE TEDDY" WAITS FOR "THE UNKNOWN."

May 26, 1903, President of the United States, Teddy Roosevelt visits Wallace, Idaho at sports field, southwest part of town, corner of River and 3rd Streets. *(Courtesy Butch Jacobson)*

May 26, 1903, President of the United States, Teddy Rosevelt visits Wallace, Idaho, photo shows his procession travelling down 6th Street. *(Courtesy Butch Jacobson)*

A crowd greeted President Theodore Roosevelt at Third and River Streets in spite of inclement weather on May 26, 1903. While Roosevelt was in town, he was a guest at the home of Al and May Hutton at 221 Pine. The beautiful school with the Moorish tower (center) was built in 1892. The large annex on the right was added in 1901. This is the school built in 1892, with the Moorish tower. *(Butch Jacobson)*

The Mighty Hercules Contribution to Spokane: The Paulsen Buildings

One of the most colorful stories involving Inland Northwest mining history, the Coeur d'Alene Mining District, and the building of Spokane was the discovery of North Idaho's richest silver vein during the mining boom days of the Coeur d'Alenes. That discovery was at the Hercules Mine and the discoverer was August Paulsen, a new resident of Wallace.

The mine had been staked in 1889 by Harry L. Day and Fred Harper. It was a mile and a half north of Burke, Idaho. Harry, his father, and other family members, had come to the Coeur d'Alenes in 1886, where they engaged in the stock and the dairy business. But they also turned their attention to prospecting.

After staking the mine, Day sought out men he had come to know and trust to enter into a partnership with him. Among them were railroad engineer Levi W. "Al" Hutton, who bought 3/32 interest in the mine for $880, August Paulsen bought 1/4th interest for $850. Among Day's partners were some other local businessmen – Paulsen's employer, a butcher, a barber, Harry Orchard had 1/16th interest but got rid of it to pay a debt to storekeeper Dan Cardoner in 1898, Shortly after he obtained his interest. (Targeted for his role in quelling a miners' strike in 1899, former Idaho governor Frank Steunenberg was mangled by a powerful bomb placed at the entrance top his gate by Orchard. The bomb was triggered when he opened the gate to his home in Caldwell, Idaho. He died shortly afterwards in his own bed.

August Paulsen, age 20. *(Courtesy Joel Moore)*

Having little capital with which to develop the mine, the partners had to compensate by doing the manual labor themselves. Most of Paulsen's contribution to the Hercules development was in the form of labor. He was the only one of the partners who worked five to six days a week in the mine; the rest had to work around other employment obligations and usually worked at the mine on the weekends. For 12 years, the partners followed a small vein of ore in search of a major discovery.

During the early part of 1901, while working at the mine alone, Paulsen was visited by a dynamite salesman. While at the site, the salesman requested a tour of the mine. As they walked to the face of where Paulsen had been working, the salesman noticed a slight mineralized rock formation on the side of the main drift. He suggested Paulsen stop where he was working and follow the ore showing he had just noticed. Paulsen took his advice, and on June 1, 1901, he discovered the vein of silver that turned the owners of the Hercules Mine into millionaires.

The history of the Hercules is especially unique among the major mines in the Coeur d'Alenes because, typically, the mines were developed by large companies with a lot of capital to back them. The prospectors who discovered the mines usually did not have the kind of capital required to develop a mine and sold their mineral rights early on in the process. However, with the Hercules, the same group of people responsible for the discovery of the mineral veins were also the ones to develop and operate the mine.

Of the original partners, Harry Day and his family became one of the most powerful families in the

The Paulsen Dental and Medical Building (left) nearing completion in 1929 and the adjacent Paulsen Building, completed in 1908. The two buildings are on the south side of Riverside and stretch from Washington to Stevens. The Paulsen Building was designed by Dow and Hubbell (John K. Dow and Clarence Z. Hubbell) and the Dental and Medical Building by Gustav A. Pehrson. Both buildings are steel-frame brick structures with glazed terra-cotta trim, produced by the Washington Brick and Lime Company. The general contractors were Frederick Phair on the first building and Rounds-Clist Company Inc. on the second one. The 11-story brick Paulsen Building was the tallest in Spokane until the 15-story Old National Bank was built across the street in 1910. *(Libby Studio photo courtesy Joel Moore, Paulsens' grandson)*

Coeur d'Alenes. Their influence extended over a period of more than 30 years. They continued to expand their mining interests, purchasing and operating numerous other mines in the area. Three of the other original partners (Paulsen, Hutton, and Frank M. Rothrock) moved to Spokane and invested in the future of the city.

The Paulsens

August "Gus" Paulsen came to the United States from his native Denmark in 1888, at the age of 17. He eventually made his way to Spokane in 1892. He worked on a dairy farm until he saved enough money to fulfill his ambition of moving to the Coeur d'Alene Mining District, which was still in the discovery period. After arriving in the Coeur d'Alenes, he again found employment on a dairy farm.

While living in the mining district, he met and married Myrtle White, originally from Colfax. After striking it rich, they subsequently moved their family to Spokane. The Paulsens were among Spokane's early self-made pioneer families who returned their wealth to the community in many ways.

Their name is most recognizable because of the two major downtown structures that bear the Paulsen name. But, in 1915, August Paulsen also built the Clemmer Theater, which is now one of Spokane's premier performing arts theaters. It was built in the early days of motion pictures, when theaters were designed to look like opera houses and seat large numbers. Howard S. Clemmer operated the Clemmer, which later became the Audian, then the State Theater, the name when it closed as a movie theater in 1985. It was subsequently renovated by Paul Sandifur Jr. of Metropolitan Mortgage, and reopened as the Metropolitan Performing Arts Center, simply known as The Met. It was subsequently purchased by Mitch Silver, and became the Bing Crosby Theater in 2006.

August Paulsen's wife, Myrtle, was a remarkable individual in her own right. Construction had recently commenced on the Paulsen Dental and Medical Building when August died on March 11, 1927, but Myrtle carried the plan out to its completion in 1929. During her lifetime, she devoted countless hours to various charitable organizations. Among them, her favorite was the Spokane Chapter of the American Red Cross. She began her volunteer service with them in 1918 and served as Spokane director and chairman of volunteer services through two world wars, from 1918 to 1947. She also provided free office space in the Paulsen Buildings for the Spokane chapter. Myrtle died on January 5, 1959. She donated their home at 245 East 13th Avenue to the Cathedral of St. John. It is presently the headquarters for the Episcopal Diocese of Spokane.

The August Paulsen Family

August Paulsen was born near Copenhagen, Denmark, July 29, 1871. He received some schooling in his native land, but was largely self-educated. In 1888, at the age of 17, he had saved enough money to come to America. He spent his first winter in San Francisco attending business college. For several years, he worked in the woods of northern Michigan and on lumber vessels. In 1892, he came to Spokane where he worked on a dairy farm for $6.00 a month. During that time he saved enough money to

August Paulsen with son.

August Paulsen's home

The north side of the Paulsen home. The family donated this home to the Cathedral of St. John.

fulfill his ambition of moving to the Coeur d'Alene Mining District, which was still in the discovery period. Two years later he bought a bicycle and rode the Mullan Road to Wallace, Idaho, where he again found employment on a dairy farm (Markwell's Dairy is now part of Silverton). He later became a partner in the Hercules mine. August met Myrtle White in Wallace when her sister Mrs. Anna Conners introduced them. He saw Myrtle again when she visited the Hercules Mine with the sister of Jerome Day's secretary, the day after Paulsen's discovery of the main Hercules vein. On September 15, 1902, August and Myrtle were married.

In the early 1900s, the Paulsens moved to Spokane. They had four children: Clarence, Howard, Pauline, and Frances. During the 12 years August Paulsen worked at the Hercules, he accumulated fatal amounts of rock dust in his lungs, which plagued his health throughout his life. Shortly before Christmas in 1926, August and his wife traveled to California. He hoped that a warmer climate and being attended by a respiratory physician might cure his lung problem. On March 11, 1927, August passed away at the age of 57 from silicosis caused from breathing rock dust.

Myrtle (White) Paulsen

A *Wallace Press* description of the Burke, Idaho in 1888 stated:

> *The town of Burke has two mines in operation, one concentrator (Tiger), seventeen saloons, four general stores, one beer hall, two boarding houses, two hardware stores, one fruit and confectionery store, one butcher shop, one livery stable, one lawyer, one physician, one furniture store, one baker's shop, about 800 inhabitants a large visiting element.*

Some of the principal Hercules Mine owners. August "Gus" Paulsen is on the wood pile holding his kitty. He was the person that broke into the ore. Others are (L to R): Unidentified (sources vary on this man's identity), Emma Markwell, H.F. Samuels, May Hutton, Jerome Day, Miss Hedin (possibly Hadeen), Myrtle White (later Paulsen) and Levi "Al" Hutton. The Hercules was developed and operated by the original group of owners, a unique situation in the district. *(Courtesy Joel Moore, Paulson's grandson)*

The Hercules Mine at Burke, circa 1930, after its closure in 1925. It produced over $43 million in silver and lead deposits. The "Mighty Hercules," as it was often called, was one of the richest silver discoveries ever made in Idaho. *(Courtesy Butch Jacobson)*

Historic President Roosevelt Cover Photo Information

By John Amonson, Wallace historian and former director of the Wallace Mining Museum

Aspiring for upward mobility in the job market is the name of the game for almost everyone wanting to make life better.

It is not unheard of, however, for professionals or executives to seek out a less-challenging occupation to become an artisan or laborer just because they prefer to work with their hands. For others it might be the desire for a location change or romantic opportunity.

One of the more uncommon such moves is that of Frank Hess. Frank was a lawyer. Not long after passing his bar exam, he moved to the upstart mining town of Wallace, Idaho. The opportunity for litigation was everywhere. Mining claim boundary disputes, extralateral deposit infringements and similar activity provided plenty of opportunity for meaningful and financially rewarding work.

He would eventually set up shop about a block away from the residence of Harry White, who was himself a co-defendant with Wyatt Earp in a claim jumping case in the Eagle-Murray area in 1884. While there was the opportunity to make a productive career in the courtroom, Frank had problems with his conscience. By his own admission, it was not the activity in court that was giving him second thoughts but the back room and bar room "deals" that caused him to turn his back on law and seek out his next move which was to the photographic office of Thomas N. Barnard and his niece Nellie Stockbridge.

Barnard was in the process of turning over a major portion of the business to Nellie, which opened the door for the need for an additional photographer. Frank Hess landed the job. Some of his most rewarding work in that field came about as a result of the efforts of Weldon Heyburn, a former mine owner and successful candidate to become a U.S. Senator. Heyburn was aware of Theodore Roosevelt's plan to visit the western states in May of 1903, and arranged to have him make a stop in Wallace even before his Spokane visit. The arrival date was set for May 26. The town went all out in its preparations. $5,000, a lot of money in those days, were spent on flags and bunting.

Although many individuals assumed that Nellie Stockbridge herself would be capturing the events of the day, it was actually Frank Hess behind the camera for the procession through town. Those images are still treasures after more than a century.

Frank continued working for Nellie for a few more years. In 1907 alone, he ventured out to almost all of the major operating mines and populated areas to document what the county looked like then. Frank was still working for Nellie both times that Clarence Darrow came to Wallace for the Steve Adams coroner's inquest and subsequent trial, but I am unable to link any photo of him to Frank Hess.

However, photography was not to be Frank's lifelong profession. By 1911, Frank had set up shop as a journeyman plumber, near where the post office building now sits on Cedar Street. He continued in that profession until his retirement.

Frank and his wife were neighbors of mine on the South Hill in Wallace as I was growing up. Except for selling them freshly picked huckleberries, or delivering his newspapers, I didn't have any in-depth conversations with him. I certainly regret that now.

There were two movies filmed in Wallace.
Heaven's Gate and Dante's Peak

Information cited from Wikipedea: Heaven's Gate is a 1980 American epic Western film written and directed by Michael Cimino. Loosely based on the Johnson County War, it portrays a fictional dispute between land barons and European immigrants in Wyoming in the 1890s. The film features an ensemble cast, including Kris Kristofferson, Christopher Walken, Isabelle Huppert, Jeff Bridges, John Hurt, Sam Waterston, Brad Dourif, Joseph Cotton, Geoffrey Lewis, David Mansfield, Richard Masur, Terry O'Quinn, Mickey Rourke, William Dafoe, and Nicholas Woodeson, the last two in their first film roles. It is generally considered one of the biggest box office bombs of all time, and was initially described as one of the worst films ever made.

There were major setbacks in the film's production due to cost and time overruns, negative press (including allegations of animal abuse on-set), and rumors about Cimino's allegedly overbearing directorial style; the film as a result, opened to poor reviews, earning only $3.5 million domestically (from an estimated $44 million budget), eventually causing its parent studio, United Artists, to collapse, and effectively destroying the reputation of its director, Cimino, previously a rising Hollywood director from the success of his 1978 film The Deer Hunter, winner of Academy Awards for Best Picture and Best Director in 1979. Cimino had an expensive and ambitious vision for the film, pushing it nearly four times over its planned budget. Its resulting financial problems and United Artists consequent demise led to a move away from the brief 1970s period of director-driven film production in the American film industry, back towards greater studio control of films, as had been predominant in Hollywood until the late 1960s.

A sceen set for Heavens Gate. Recently the assessment of Heaven's Gate has become more positive, with some critics now describing it as a "modern masterpiece."

Dante's Peak

Information cited from Wikipedea: Dante's Peak is a 1997 American dramatic disaster thriller film directed by Roger Donaldson. Starring Pierce Brosnan, Linda Hamilton, Charles Hallahan, Elizabeth Hoffman, Jamie Renée Smith, Jeremy Foley, and Grant Heslov. The film was set in the fictional town of Dante's Peak wherein the town must survive the volcano's eruption and its dangers. It was released on February 7, 1997, under the production of Sony, Universal Pictures, and Pacific Western. Principal Photography began on May 6, 1996. The film was shot on location in Wallace, Idaho, with a large hill just southwest of the town digitally altered to look like a volcano. Many scenes involving townspeople, including the initial award ceremony, the pioneer days festival, and the gymnasium scene were shot using the actual citizens of Wallace as extras. Many of the disaster evacuation scenes that did not involve stunts and other dangerous moments also featured citizens of Wallace; dangerous stunts were filmed using Hollywood extras. Mount St. Helens also makes an appearance at the very end of the movie; during the start of the closing credit crawl, the scene shows an image of a destroyed Dante's Peak community with the camera shot moving out to show a wider scene of disaster, and then showing what remains of the volcano itself. The volcano that remains is actually an image of Mount St. Helens, taken from news footage just after the May 18, 1980, eruption.

Archie Hulsizer, Wallace, and Dante's Peak

From left: Linda Hamilton, Debra Mikesell (former mayor of Wallace), Pierce Brosman, Archie Hulsizer, and Bill Dire. *(Photo Courtesy Archie Hulsizer collection)*

Archie was born in Spokane in Saint Lukes Hospital, in 1921. He is a retired bank executive from Wallace, with many accomplishments in his 96 years. Archie's father died in a truck-train accident in 1921, when he was seven weeks old. His mother remarried Authur Robas in 1924.

Archie moved to Wallace in 1939, when he got the job at the First National Bank of Wallace, making $80 a month. In 1939, $1.00 had the same buying power as $17.25 in 2017. Archie and his beautiful wife, LoRayne, lived in the same home for 64 years since they first came to Wallace. LoRayne and Archie first met when she was 16 and he 18. Prior to her death, they had almost 73 wonderful years together.

Archie was an accomplished musician. He played clarinet, alto sax, and piano. In high school he was in band, orchestra, pep band and dance band. He played at the Spokane Country Club, the Hayden Lake Country Club, and the Davenport Hotel. He also played in Lewiston and Harrison in 1938-39, and all around the area. During the Depression, he played for dances for 50 cents a night. His early years were spent in Spokane. During that time, he had the honor of being the fastest typist in his high school class.

Following Pearl Harbor, Archie enlisted in the Air Force, spending 43 months in the service. He was discharged on Valentines Day 1946. His last assignments were Eniwetok and Iwo Jima, where he was assigned to a P-47 fighter group.

Archie started working at First National Bank of Wallace in July 1940, where he spent 70 years. It was the first nationally organized bank in Shoshone Coun-

The Republic P-47 Thunderbolt was a World War II era fighter aircraft produced by the United States between 1941–1945. Its primary armament was eight .50-caliber machine guns. In the fighter-bomber ground-attack role it could carry five-inch rockets or a bomb load of 2,500 pounds *(Public domain)*

ty, which included half of Idaho. He worked in the building until 2010, and retired as senior vice president in 1986 at the age of 65. Following his retirement he remained on the board of directors until the bank was sold to the First Security Bank of Idaho in 1991. When it was later sold to Wells Fargo, he was asked to be on the northern region advisory board of directors for them.

Banking was completely different in those days. Wallace was a small town, you knew everyone in the community and looked out for their needs. People would rather die before they wouldn't pay you back. At one time, a bank examiner asked Archie where the credit files were. He didn't know what he was talking about because he'd never seen a credit file. Back then, you loaned people money and they paid you back. Archie was also very active in his community. He was on the board of the Northern Pacific Depot Foundation, being influential in getting the depot moved across the creek during the I-90 expansion plan.

He was also on the board of the Wallace District Mining Museum for years. Many donations of historical objects, like the original 1892 minutes from the bank, and a pistol from one of the teller cages were among them. He was also a trustee of the Wallace Public Library and director of the Wallace Public Library Foundation. He was also school district treasurer for 40 years. He belonged to the Congregational Church and directed the choir there for over 25 years.

He's also the oldest living past exalted ruler of the Elks Lodge, and still an active member there. He originally joined the Shoshone Masonic Lodge in 1946. He's been a member of the American Legion for over 64 years and a lifetime member of the Masons and VFW (Veterans of Foreign Wars).

He was also chairman of the Wallace Centennial in 1984, a one-year celebration. He was the mayor of Wallace for about three years and spent 17 years on the city council before that, being recruited for those jobs each time.

During the filming of "Dante's Peak," he was mayor. At that time, Linda Hamilton once gave him a kiss on the right cheek. He didn't want to wash his face for days after that.

He was president of Slippery Gulch for 33 years - they do old-time events, like a show where maybe 10 guys in tutus will put on a swan dance. This was a fund raising event for community projects

He was also president of the Callahan Zeller Foundation. They meet once a year and give away any money they've earned.

On July 18, 1990, Gov. Cecil Andrus signaled a proclamation proclaiming "Archie Hulsizer Day in Idaho."

Archie spent nine years umpiring Little League, Babe Ruth, and American Legion baseball games once or twice a week. He umpired local games as well as District, State and Regional tournaments.

He was the grand marshal for two different parades.

Wallace Mayor Herman Rossi Gets Arrested for murder in 1916 Shoots Clarence "Gabe" Dahlquist in the Lobby of Samuel's Hotel

The Samuels Hotel located at the corner of Cedar and Seventh, had been completed and opened on the first of May, 1908, only a year after the completion of Shoshone County's stately courthouse a block away. (Inset Samuels) The hotel was demolished in 1975, and later became the Harry Magnuson Park. *(Courtesy Butch Jacobson, Barnard/Stockbridge photo)*

The Samuels Hotel, was built by Henry Floyd Samuels, a lawyer whose career, by 1908, had expanded to include a series of profitable interests in local mining properties. Samuels arrived in Wallace in 1895. From 1896 to 1898, he served as the city attorney for Wallace. In the later part of 1896, he was elected as the first county attorney for Shoshone County, a position he occupied for two years.

The Samuels Hotel was considered to be the largest and finest hotel in Idaho in its day. George M. Teale, a writer for the *Overland Monthly* wrote the following about the Sammuels Hotel in the November, 1908 edition of that magazine:

> *There are many beautiful homes here with every modern convenience," "...but the pride of the city, and of the whole district for that matter, is the new 'Samuels Hotel,' a modern five-story brick structure that would be a source of pride to a city of 100,000 people" According to Teale's account, the hotel offered more than 150 rooms, including 25 suites. Guests found there the height of modernity in hotel accommodations. "Each room," wrote Teale, "has hot and cold water, electric light, steam heat, long distance telephone, open nickel plumbing, brass beds, and, in fact, the best of furnishings in every respect." Handsomely outfitted office space occupied the ground floor. Guests could get a haircut and a shave in the hotel's barbershop, refreshments at its "Metals Bar," and "a passenger elevator" took them to upper floors. The hotel's elegant café accorded diners flowers, cut glass, and excellent service.*

Teale's article reported that Samuels had invested more than $250,000 in this building and outfitting his hotel – a sum equivalent to about $5.5M in 2016 dollars. The hotel, Teale added, was "very popular, and nearly always full."

Samuels later turned his attention to mining and was one of the original owners of the Hercules, Stewart, and Success mines. He later left the area, but returned to Wallace with a Master of Law degree, granted by Columbian University. He also developed an effective process for separating zinc from lead and silver, thus earning himself the honorary title, according to one source, of "father of the zinc industry of Idaho."

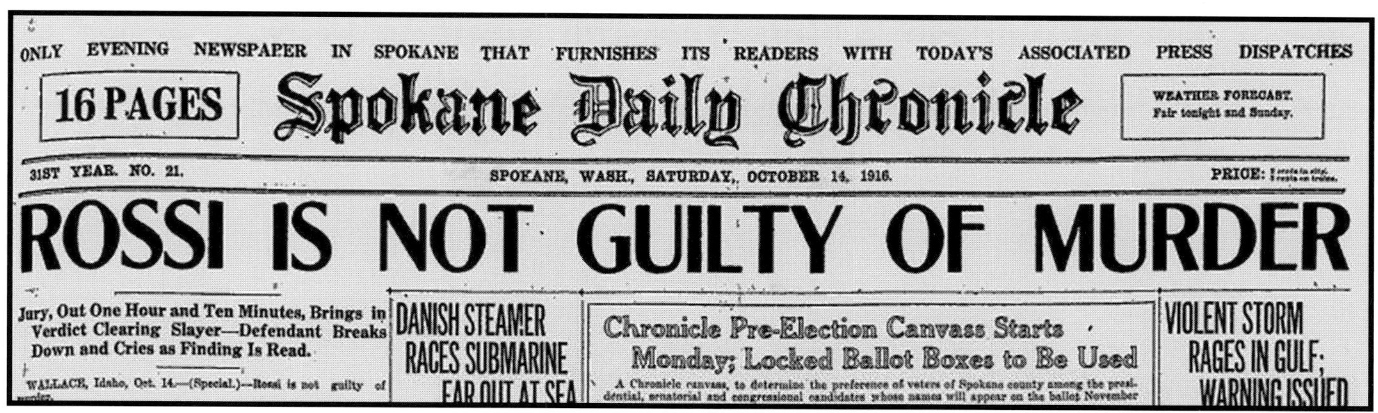

ONLY EVENING NEWSPAPER IN SPOKANE THAT FURNISHES ITS READERS WITH TODAY'S ASSOCIATED PRESS DISPATCHES

16 PAGES

Spokane Daily Chronicle

WEATHER FORECAST.
Fair tonight and Sunday.

31ST YEAR. NO. 21. SPOKANE, WASH., SATURDAY, OCTOBER 14, 1916. PRICE:

ROSSI IS NOT GUILTY OF MURDER

Jury, Out One Hour and Ten Minutes, Brings in Verdict Clearing Slayer—Defendant Breaks Down and Cries as Finding Is Read.

WALLACE, Idaho, Oct. 14.—(Special.)—Rossi is not guilty of murder.

DANISH STEAMER RACES SUBMARINE FAR OUT AT SEA

Chronicle Pre-Election Canvass Starts Monday; Locked Ballot Boxes to Be Used

A Chronicle canvass, to determine the preference of voters of Spokane county among the presidential, senatorial and congressional candidates whose names will appear on the ballot November

VIOLENT STORM RAGES IN GULF; WARNING ISSUED

The Saturday October 14 Spokane Daily Chronicle for October 14, 1916, *(Public domain)*

On June 30, 1916, Herman Rossi returned to Wallace Idaho to find his wife drunk in her bedroom and telling him a story that she had just committed a number of immoral acts with her lover, Clarence "Gabe" Dahlquist. Upon learning this, Rossi hurried to the downtown Samuels Hotel, where he knew that Dahlquist had a room. As he entered, he found Dahquist sitting in the lobby. This encounter began sometime between 6:30 and 7:30 p.m. As Rossi approached Dahlquist he struck him on the head at least three times with a .38 caliber revolver. A brief scuffle between the two men then occurred, after which Dahlquist "got up and ran."

"As he was crossing the hotel lobby," testified J. L. Jones, "Rossi raised his gun and fired." The bullet pierced Dahlquist's lung. Now gravely wounded, "Dahlquist veered in his course and crawled under a desk behind the cigar case." Rossi pursued him and was about to shoot him a second time when George Baxter, a clerk at the hotel, and Mrs. Laura Stone, behind the register, urged him not to. Baxter testified he implored Rossi: "For God's sake, Herman, don't shoot." Stone pleaded to Rossi that he refrain from doing something he would always regret.

MRS. MABEL ROSSI

Both interveners noted that Rossi seemed to regain control of himself when they spoke to him. Rossi, for his part, warned Dahlquist that he would kill him "if he did not leave town within ten minutes." Dahlquist responded to the effect that he would do so if he could find a car. On his way out of the hotel, Rossi stopped momentarily to offer an apology to hotel manager Dave Johnson, "for having this thing happen in his house." Rossi then went to his attorney's office where he was arrested, taken to the police station, booked, and released on $10,000 bail put up by some of the town's most prominent citizens.

During the funeral of the victim, Dahguist, the people of Wallace turned out in such large numbers that it was the best-attended funeral event held up to that time in Wallace.

It should be noted that Herman Rossi's (second) wife, who was fifteen years younger, had a drinking problem which became worse during their marriage. Rossi made attempts to cure her, but nothing seemed to work.

Three and a half months later, Rossi was unanimously found not guilty by reason of temporary insanity by a hometown jury after less than twenty minutes deliberation. Interestingly, the judge was a fellow Elk and Mason and a co-director with Rossi in the Amazon Dixie Mine. There was never any question that Rossi shot the man, but there was a general community feeling that he had been grievously wronged and had made just retribution.

The following year and fully exonerated, Rossi divorced Mable and married Bernice Johnson of Boise in May.

Mayor Rossi Gets Arrested Again in 1929

The Herman Rossi Building on Bank Street. *(Courtesy Butch Jacobson)*

Herman J. Rossi was one of Wallace's more successful and colorful men, and his investments were numerous. One of the current mines which is still active is the Golden Chest Mine in Murray. When the original owners of the Golden Chest lost interest, by 1936, Herman J. Rossi, the colorful four-time mayor of Wallace, insurance agent, and mine owner, become the company's president. However, his death on March 12, 1937, removed the last vestige of interest from this company. Rossi also had numerous other mining interests.

Though Shoshone County was not much different than countless other regions in its violation of the Prohibition Act, it made headlines nationwide after a series of raids and numerous arrests in Wallace, Mullan, and Kellogg in 1929. They captured attention due to the large number of county and city public officials who were arrested and accused of widespread conspiracies to condone, protect, and illegally benefit from illicit liquor trade. During their fervid cleanup, federal undercover agents arrested nearly two hundred residents, including Shoshone County Sheriff Rene E. Weniger; sheriff deputies Charles Bloom and Albert Chapman; Wallace Mayor Herman J. Rossi; Shoshone County Assessor William H. Herrick (also a former Wallace mayor); and high-ranking police officers, including the chiefs, of Wallace and Mullan.

Most of the arrested were found guilty and handed federal prison sentences, along with hefty fines. However, within the next two years, a large percentage of the cases were overturned during the appeals process. Allegations of organized conspiracy could not be adequately proven, and the city and county officials had not profited personally by their acts. Lawyers successfully argued they had operated "for the good" of their county or communities. Their acts involved the collection of license fees (essentially amounting to fines) from saloons (thinly disguised as soft-drink parlors) and other illicit establishments, such as gambling halls and houses of prostitution. Some were also known to have accepted bribe money to look the other way (an offense not so readily viewed as "for the good of the community").

The officials involved justified their actions by the effectiveness of the fines system, which helped to cover municipal projects and defray some of the additional costs imposed by the Volstead Act. Prohibition had stamped out the considerable revenue the legitimate liquor licenses had previously produced. For elected officials to interfere with this new source of revenue, or to take a hard line in enforcing the bans on liquor trade, would likely have been political suicide.

A 1908 sketch of Herman J. Rossi, well-known, four term, mayor, insurance executive, and mine owner from Wallace. Rossi was president of Jack Waite Consolidated Mining Company from the time of its formation, on January 30, 1928, until it was reorganized in August 1930 as the Jack Waite Mining Company (John Duthie's interests). *(From* Inland Emperors *by William E. Kinnear)*

Installation of the First Telephones in Wallace

First telephone installers in Wallace, circa 1888. Thomas Elsom (front row, right) supervised the installation of the lines into the mining district. By November 1888, every mining town in the Coeur d'Alenes was connected by telephone. Elsom kept meticulous diaries. The entry of January 15, 1888 (while in the midst of stringing the telephone lines from Spokane into the mining district) read: "At Mission. Thermometer frozen at 40 below. I am a fool if I am here a year from now." The following year, he took charge of the Wallace office. Elsom was in Wallace when he received word of Spokane's disastrous fire.

(Courtesy Dean Ladd & Larry Elsom)

The Four Waterways Flowing Into and Out of Wallace

1) Canyon Creek flows for six miles from Burke Canyon into Wallace, where it joins the South Fork of the Coeur d'Alene River. Driving up the canyon, just outside of Wallace, there are numerous remains of mines and towns, that existed during the heyday of lead/silver mining. The pollution being dumped into Canyon Creek was a major ecological problem.

Burke, which is up Canyon Creek, began in 1884, with the discovery of rich ore. Mines and mills began filling the surrounding hillsides. The largest mine in the early days was the Hercules Mine. Although it was filed on by Harry Day on August 24, 1889, it wasn't until 1901 that the richest silver vein was discovered by August Paulson, then a partner of Harry Days, with a one fourth interest in the mine, then called the Hercules. It operated until 1925.

Colonel Wallace was the founder of the Oreornogo, which got its start up Burke Canyon. Due to the narrowness of the hundred-yard wide canyon, the railroad ran through the middle of the town. Most towns had main streets. However, Burke had a railroad, and due to its narrowness, most of Burke's buildings

Canyon Creek Road up Burke Canyon. All the canyon's towns were a pollution disaster, with outhouses and other buildings built on stilts above the creek. The area's sewage flowing into the creek and on to the city of Wallace, then to the South Fork and on. Photo 1928. *(Courtesy Butch Jacobson collection)*

The main street of Burke in 1904. Built along Canyon Creek in a valley so narrow that Burke's main street was shared by two rail lines. Burke had a unique distinction of being the only mining town known to have a railroad before it had a wagon road. D.C. Corbin built a narrow-gauge railroad from Old Mission on the Coeur d'Alene River to Burke in 1887. In 1906 the Northern Pacific built a spur to the Hercules Mine loading platform at the east end of Burke, constructing its line through the four-story Tiger Hotel. There were actually two Tiger Hotels. The first one was built in 1888, but burned sometime around 1896, It was located next to the north hillside quite a distance from the tracks. the newer hotel was built a few hundred feet farther up the canyon. In 1906 the Northern Pacific Railroad was extended through the hotel to reach the shipping docks which were further up the canyon. *(Courtesy Dean Ladd & Larry Elsom)*

crowded up against the railroad, or were built over it. In 1884, Glidden bonded the Tiger mine with John M. Burke (First lode discovered in the CDA's). Soon after, Glidden moved his family to Spokane and began operating his mining interest out of Spokane, he became one of the largest shareholders of the Tiger Mine. He was also one of the stockholders after its consolidation with the Poorman.

About four years prior to his death, he sold his interests in that property to Charles Sweeny and concentrated his energies largely upon financial interests in Spokane. In 1890, he established the Old National Bank, which later became one of the largest banking institutions of the northwest. Stephen Glidden married Sue Garret from Spokane and together they had seven children. Glidden was among those who made their fortunes in the Coeur d'Alenes. He was born in Northfield, New Hampshire, in 1828, and passed away at Los Angles, California, on March 17, 1903.

In 1890, a second railroad was later built, which also ran up the same street. It also eventually went through the Tiger's hotel's lobby. Passengers got on and off the train in the hotel lobby. It was written in *Ripley's Believe It or Not,* that the town was so narrow, merchants cranked their storefront awnings up so the trains wouldn't knock them off as they passed by.

2) The South Fork of the Coeur d'Alene River, flows from the east for 37 miles until it meets the North Fork of the Coeur'd'Alene River near Enaville. As it flows through the city of Wallace it would pick up the pollution from the other creeks flowing into the town, and pass it on with the flow into Coeur d'Alene Lake. As they flow westbound out of Coeur d'Alene Lake, those waters are now called the Spokane River.

3) The East Fork of Nine Mile Creek is one of the most polluted tributaries of the Coeur d'Alene River. The Interstate-Callahan Mine, in Nine Mile Canyon, ushered in an era of industrial mining that extracted enormous wealth from the canyon, but in turn, created a vast amount of mining pollution in Idaho's Silver Valley.

Small company towns that sprang up around the Interstate-Callahan and two other large mines in Nine Mile Canyon, are now gone. Cave-ins have closed the mine portals. Aerial trams that once hauled ore from the mines to processing facilities exist only in historical photographs. Enormous piles of waste rock still remain.

4) Placer Creek flows into the west side of Wallace from the south, travels almost the distance of Wallace and enters the South Fork of the Coeur d'Alene River across from Second Street.

Placer Creek was an area of little mining activity. However, its major historically recorded event happened during the summer of 1910. By August, there were approximately 1,400 wildfires burning out of control in northern Idaho, northeastern Washington, and northwestern Montana. On August, 20, some of these wildfires blew up into huge fire storms, threatening a number of rural communities and trapping a number of fire fighting crews near Wallace. One of those crews was led by Ed Pulaski.

Pulaski and his crew were fighting the fire in an area around the St. Joe Divide and Lake Elsie region approximately 10 miles southwest of Wallace. As the fire appeared to be life-threatening, Pulaski ordered his forty-five men to head toward Wallace. As they retreated toward Wallace, a second fire blocked their way. They were now surrounded by fire. Fortunately, Pulaski knew the area and was able to lead his men to an old abandoned mine tunnel. The crew (along with two horses) entered the mine opening near the west fork of Placer Creek about two miles above its confluence with the main channel just as the fire engulfed the area.

The mine was actually a short prospecting tunnel that had been abandoned because no ore was found there. The mine opening was 6 feet high and 5 feet wide. The entrance had been timbered with logs. The mine tunnel extended 250 feet into the hillside.

As the fire raged outside, the heat caused the cold air in the tunnel to rush out, replacing it with hot air and smoke. Pulaski ordered the men to lie down on the floor of the mine where there was still some breathable air. As the fire closed in around the tunnel entrance, the timber beams at the mouth of the tunnel began to burn. Pulaski carried water in his hat from pools on the mine floor to the entrance, in an attempt to douse the flames. In the process, he was badly burned. Eventually, Pulaski fell unconscious, as did all his men.

The following morning the fire had passed by them. One of his crew had been killed by a falling snag on the way to the mine; five others had died from smoke inhalation during the night, the two horses were so badly injured they had to be shot. Thirty-nine had survived. As the fire in the surrounding forest died down, Pulaski and his crew followed Placer Creek to safety in Wallace.

In two days, the Great Fire of 1910 consumed 3,000,000 acres of forest. The six men lost in or near Pulaski's tunnel were among 78 firefighters killed by the fire. There were also seven civilians who died in the fire; bringing the total loss of life to 85 people. Because of its association with Ed Pulaski and the Great Fire of 1910, the Pulaski Tunnel and fire escape route were listed on the National Register of Historic Places in 1984. The famous fire fighting tool still used to combat wildfires today was developed by Ed Pulaski, and called a Pulaski.

The Pulaski Story as it relates to Wallace, Idaho
Quoted from the *Coeur d'Alene Gold Rush and Its Lasting Legacy*

This is one of numerous Forest Service photos showing the aftermath of the murderous 1910 fire. This scene was representative of the appearance of three million acres of blackened terrain that stretched across northwestern Montana into northeastern Washington, cutting a wide swath through northern Idaho. In this photo, a crew is cutting a trail through the downed timber. The burned trees that were still standing were extremely hazardous, especially if a wind whipped up, because they toppled easily. Today, burnt snags from this fire still stand as a reminder of the terrifying event, which remained forever etched in the memories of those who experienced it. *(USFS Ranger Joseph B. Halm photo, Museum of North Idaho, Fi-1-25)*

Ranger Edward Pulaski and his white horse, a familiar sight around Wallace, passing through a cut in the remains of a snow slide on Placer Creek. *(William W. Morris collection, courtesy Butch Jacobson)*

The following account of Edward Pulaski, and the related photos, was taken from Tony and Suzanne Bamonte's *The Coeur d'Alenes Gold Rush and Its Lasting Legacy*:

Among those the government turned its back on during his terrible time of need was Edward C. Pulaski, the assistant USFS ranger from Wallace, who today is revered as the legendary hero of the 1910 fire storm. Pulaski's crew of firefighters were spread across the mountains of the St. Joe Divide, in the vicinity of Lake Elsie, when the winds picked up. Fortunately, Pulaski was familiar with the forests around Wallace, and his past experiences and skills had prepared him for wilderness survival.

Just before they were overtaken by fire, Pulaski rounded up nearly fifty firefighters (accounts vary) and led them to an abandoned mine tunnel on the West Fork of Placer Creek. At the risk of his own life, he stood guard at the entrance to prevent anyone from leaving, holding at gunpoint some who tried to escape. Tragically, during that dreadful night, five men inside the mine died, and another was overcome a short distance from the entrance. However, though suffering from smoke inhalation and other injuries, the rest of the men survived because of Pulaski.

Pulaski's hands, head, and eyes were severely burned, causing him some permanent vision loss and unrelenting pain until his death in 1931. His throat and lungs were also damaged. His legend grew as memories of the fire faded and survivors told their stories of his heroism. But he became widely known for another reason: his design of the Pulaski, which became the most

Ranger William Morris and Emma Pulaski, wife of Edward Pulaski, circa 1910. The Pulaskis were married in 1900 and were the parents of a daughter named Elsie. *(William W. Morris collection, courtesy Butch Jacobson)*

Legendary hero Edward Pulaski, assistant USFS ranger from Wallace, after recovering from the 1910 fire. *(Museum of North Idaho, FS-8-28)*

commonly used hand tool in fighting forest fires. It is a combination of an axe and a grub hoe (or adze), which allows a firefighter to either chop or dig with a single tool.

Pulaski continued as district ranger of the Wallace region until 1929. He lobbied the government for many years to erect a monument or at least a marker to honor the firefighters, meanwhile continuing to personally tend to the scattered graves of his fallen comrades. Finally, in 1921, the government approved a small appropriation to create a bronze plaque with the names of the firefighters who lost their lives. In 1933, a 1910 firefighters memorial was created at the Woodlawn Cemetery in St. Maries, Idaho. Unfortunately, Pulaski did not live to see it. A memorial was also eventually created to honor Pulaski, which was dedicated at the centennial commemoration of the 1910 fire. It is just south of Wallace, where he led his crew of firefighters to an abandoned mine.

An excellent story regarding Edward Pulaski was written by Becky Kramer for the *Spokesman-Review* on Aug. 17, 2010:

Forest ranger sheltered crew in former mine-shaft during 1910 fire
Pulaski's legacy alive in standard fire tool
Complete coverage
Flame and Ruin: The fires of 1910

The west fork of Placer Creek is a friendly little stream, trickling through a narrow canyon lined with ferns and cedar trees.

Follow a two-mile trail up the creek, and you come to the Pulaski Tunnel – a legendary part of the 1910 Fire story. It was here that Big Ed Pulaski ordered 45 firefighters into a mine shaft on the night of Aug. 20, and told them to lie face down.

"One man tried to make a rush outside, which would have meant certain death," Pulaski later wrote. "I drew my revolver and said, 'The next man who tries to leave the tunnel I will shoot.'"

Today, Pulaski's story is told in interpretative panels along the trail. The mine shaft itself has become a shrine to 1910 Fire buffs, who hike the trail in tribute to the capable, quick-thinking assistant Forest Service ranger who saved most of his firefighters during North America's worst fire storm.

But for decades, the tunnel's location was a mystery. It faded into obscurity after the last survivors died. In 1979, Carl Richey's supervisor at the Panhandle National Forests gave him an assignment: See if you can find the Pulaski Tunnel.

"Everyone knew the drainage, but no one knew where the tunnel was," said Richey, a retired Forest Service archeologist. "I interviewed different people who purportedly knew where it was, but I quickly learned that no one really knew anything about it."

Richey took on the project. He bushwhacked along the creek, studied old photographs taken shortly after the fire and pored over old mining claims and surveyor's reports.

Early on, Richey pieced together some important clues. The tunnel was close to the creek. It was one of five drilled by miners in the canyon. An old Forest Service sketch put the location at 1 1/2 miles up the west fork of Placer Creek.

Like the tunnel, Pulaski himself is something of a mystery. He was born in Ohio, and left school at about age 15. Letters from an adventurous uncle, who wrote vivid accounts of his life in mining camps, fueled Pulaski's own decision to head West. Pulaski was in Murray, Idaho, in 1884 for the gold rush. He worked as a packer, labored in mines and lumber camps, and picked up blacksmithing skills. By the time the Forest Service hired him as an assistant ranger in Wallace, Pulaski was 40 years old, a seasoned outdoorsman.

"Big Ed," as he was called, stood 6 feet 4 inches. Despite the jocular nickname, photos of Pulaski show a serious man, dark-haired, dignified and somewhat enigmatic.

"Mr. Pulaski is a man of most excellent judgment; conservative, thoroughly acquainted with the region, having prospected through the region for over 25 years," wrote William Weigle, supervisor of the Coeur d'Alene National Forest. "He is considered by the old timers as one of the best and safest men to be placed in charge of a crew of men in the hills."

But the lack of formal education cost Pulaksi the ability to advance in the Forest Service. He was an assistant ranger, earning $75 per month.

Richey never doubted that he'd find the tunnel. It was a matter of logical deduction. He was looking for a hole in the hillside, within six or seven feet of Placer Creek's west fork. Prospectors had built a cabin on the other side of the creek, where the canyon flattened out.

William Morris at the portal of the abandoned mine where Pulaski led his crew to safety, in 1910. *(William W. Morris collection, courtesy Butch Jacobson)*

Since there was no trail at the time, Richey drove in on an old Forest Service Road above the creek. In earlier days, mule trains packed supplies over the road from Wallace to Avery.

As he tramped through underbrush along the creek bottom, Richey scanned the hillsides. His trained archeologist eye was looking for waste rock piles, the debris left over from mining excavations. From his research, Richey knew that at least five mine shafts had been drilled in the canyon walls. "I was pretty good at visualizing things, and I walked the whole creek bottom," Richey said.

First, he found the War Eagle Mine's waste rock pile. Pulaski was actually trying to reach the War Eagle Mine as fire surrounded the crew. He knew the mine's deep shaft – it was bored 1,300 feet into the canyon wall – would provide shelter. But events unfolded unpredictably on Aug. 20, 1910, as unpredictable as the winds that whipped smaller forest fires into an inferno.

Pulaski was in charge of 200 firefighters between Wallace and Avery. On the night of Aug. 19, he'd ridden into Wallace to gather up food and first aid supplies for the crews. He had dinner with his wife, Emma, and 10-year-old daughter, Elsie. The fire would reach Wallace, he said, instructing them to take shelter on a pile of mine tailings near the family's Burke Canyon home, where the rocks would keep the fire from approaching. On the morning of Aug. 20, he headed back up the mountain. His last words to his wife: "I may never see you again."

Pulaski was with about 45 men on Striped Peak when the fire blew up. "A terrific hurricane broke out over the mountains," he later recalled. "The wind was so strong that it almost lifted men out of the saddles, and the canyons seemed to act as chimneys, through which the wind and fires swept with the roar of a thousand freight trains."

Firefighting became futile. "Boys, it's no use," Pulaski told the crew, according to one survivor's account. "We've got to dig out of here. We've got to try to make Wallace. It's our only chance." The men fled with the fire on their heels, a black bear racing alongside them. Trees exploded into flame, then toppled under 60 mph winds. One man fell by the trail, either hit by a tree, or unable to go on for other reasons. Pulaksi gave his horse to an ex-Texas Ranger, who was limping from rheumatism. As they headed down the West Fork of Placer Creek, the fire surrounded them. Pulaski contemplated taking shelter in the War Eagle Mine, but discarded the plan when he realized the mine was still too far away. Instead, he led the men to a shallow opening drilled by miners called an adit, while he looked for a larger one – the Nicholson tunnel. The men and two horses crowded into the Nicholson tunnel, but their feeling of refuge was short-lived. Mine timbers near the entrance were smoldering, sucking oxygen out of the shaft. Pulaksi wrapped wet blankets around the timbers and used his hat to scoop muddy water out of puddles on the mine floor. His hands and hair burned. The fire seared his eyes. Over the next five hours, the fire raged. Panicked men screamed, moaned, convulsed and retched. One tried to strangle another. "The tunnel became a mad house, a hellhole where five men would die," Richey wrote in a Forest Service report.

Pulaski kept the frantic men inside the tunnel at gunpoint. Finally, the tunnel was quiet. The men had passed out, some never to awaken.

Richey found the Nicholson tunnel in October 1979. Frost had stripped the leaves from the brush, making the portal easier to spot. The entrance was low, but a bit further inside, the 80-foot-long tunnel opened into a gallery-like area that rose to about 20 feet in height. Richey found an enamel coffee pot, along with two steel drill bits and an old dynamite box. About half of the tunnel was blocked by falling rock. "The timbers inside the mine were burned," Richey said. "But there's nothing that said, 'Pulaski was here.

The discovery still sent ripples of excitement through the Forest Service and the local community. Before he retired, Richey led about two dozen hikes up the creek to the mine shaft, and spoke at local chamber of commerce meetings.

Community members lobbied for nearly $300,000 in federal appropriations to pay for the trail and interpretative signs.

Today, visitors can look down at the gated portal from the trail above the creek. Charred timbers were recently installed around the opening for the 1910 Fire's anniversary. The rock opening itself is draped in ferns and other greenery. It's hard to reconcile the lushness of the site with photos taken after the fire. They show a blackened landscape, scorched trees scattered like pickup sticks.

Around 5 o'clock the next morning, the men started to stir. Pulaski's body lay motionless near the entrance. "Come outside boys, the boss is dead," one of the survivors called out. "Like hell he is," Pulaski replied. "I raised myself up and felt fresh air circulating through the mine. The men were all becoming conscious," he wrote in an account of the fire.

Five men and both horses had perished in the tunnel. The survivors hobbled painfully back to Wallace, their path strewn with smoking logs and burning debris. They were parched, but the creek water was too hot and ashy to drink.

Pulaski had to be led down the mountain. He was temporarily blinded, and spent two months in the hospital with pneumonia. He and his wife exhausted their savings paying for other firefighters' medical bills.

Today, Pulaski's story is the iconic tale of the 1910 fires. "His story just typified the cowboy: The strong, tall cowboy who saved the day," said Russ Graham, a research forester from Moscow, Idaho. "A mystique developed around the rugged outdoorsman who had firsthand experience on the land, and who used it to save his crew."

But there was no glorious ride into the sunset for Pulaski. After the fire, he was a broken, bitter man.

Pulaski returned to work with damaged lungs and blindness in one eye. A colleague, Roscoe Haines, tried to get the Forest Service to compensate Pulaski for his health problems. When that failed, Haines tricked Pulaski into submitting an account of the night in the tunnel to the Commission, hoping its Hero Fund would reward him for saving the firefighters' lives. The Carnegie commission also turned Pulaski down. In perhaps the most painful blow, the Forest Service refused to fund a granite memorial that Pulaski designed for the fallen firefighters. The $435 cost would require "an act of Congress," the Forest Service said.

"He really felt that the government abandoned him," said Jason Kirchner, an Idaho Panhandle National Forests spokesman. "He felt that the government owed these firefighters a huge debt of gratitude. Some received remuneration, but it wasn't consistent across the board. That offended him."

Pulaski retired from the Forest Service in 1929. He died two years later, of complications from the injuries he received in a severe automobile accident.

As part of the 1910 Fire Commemoration in the Silver Valley, local residents raised $45,000 for the granite firefighters' memorial. It matches Pulaski's design, and will be dedicated at noon Saturday in Wallace.

Pulaski would be pleased, Richey said.

"He wasn't an attention hog. He only wrote one little piece about the fire," Richey said. "But he would have liked the recognition for his men."

Source material for this account: "My Most Exciting Experience as a Forest Ranger," by Ed Pulaski; "Pulaski: Two Days in August, 1910," by Carl Richey; "Year of the Fires," by Stephen Pyne; "The Big Burn," by Timothy Egan.

Burke's Guido Bardelli, "Young Firpo" A Member of the World Boxing Hall of Fame Got His Start in the Coeur d'Alene Mining District

Guido Bardelli at the age of 18. He fought 25 of his fights in Wallace to packed crowds. His first Wallace fight, at the Eagles, was in 1925. *(Courtesy Bardelli collection, Barnard-Stockbridge photo)*

While driving through Wallace, Idaho, in 1924, in search of a fighter to complete a boxing card that evening, at the Morning Club in Mullan, a promoter stopped to talk with a well-built, handsome young man he recognized, standing on a street corner in town. Guido Bardelli, a 17 year old Burke High School student, listened to and accepted his offer to fight, perhaps in anticipation of the money that would help his mother; and that night knocked out his opponent in the first round – making his sensational debut as a prize fighter.

From then on he was known as Young Firpo, the Wild Bull of Burke, being named after the Argentine Italian, Luis Angel Firpo, the Wild Bull of the Pampas, who came close to knocking out heavyweight champion, Jack Dempsey, in 1923.

Wallace boxing promoters were anxious to sign Firpo as their drawing card. He stepped, for the first time, under the glaring lights of a Wallace ring, before a boisterous crowd, including a multitude of his loyal Burke fans, in the packed Eagles' Hall, on a cold winter night, in 1925.

"The Wild Bull of Burke tore loose and riddled Al Anderson of Kellogg with a series of volcanic bombardments that shook the house as well as Anderson. All Burke must have been at the fights, if the amount of cheering for Firpo is any indication," wrote a reporter, astonished together with the crowd, by the ferocity of Firpo's invincible attack, that brought him victory, to the delight of his joyous fans.

Firpo was a sensational attraction in Wallace, always drawing capacity crowds from all parts of the Coeur d'Alenes and beyond, who were thrilled by his "cyclonic" attack and "a punch like the kick of a husky male mule" and they believed they were witnessing the ascent of a future world champion. Firpo was described by Chuck Snyder, prominent Wallace promoter, as "the hardest hitting two fisted fighter that ever stepped into a Northwest ring." Although Firpo was fighting throughout the Northwest and on the Pacific Coast during the early years of his career, the majority of these fights took place in Wallace, where he won 23 of 25 fights, 14 by knockouts, and lost by two by decisions.

Having "one of the most startling records of victories ever made in the Northwest" – including nine one-round knockouts in his first 15 professional fights – promoters nationwide were eager to sign Young Firpo to headline their boxing cards. He went on to win the Pacific Coast Light Heavyweight Championship in 1933, while becoming a leading contender for the World's Light Heavyweight Championship. Firpo defended his title against some of the greatest light heavyweights in boxing history, such as Tiger Jack Fox, George Manley, Wesley "K. O." Ketchell, and John Henry Lewis.

A Butte, *Montana Standard* sports writer wrote, that the Pride of Burke and the Coeur d'Alene Mining District "has become the most sought after fighter in the West" – Firpo drawing record crowds wherever he fought.

In 1934, Firpo signed to fight Maxie Rosenbloom for the World's Light Heavyweight Championship, in Portland. However, a dispute over the gate receipts led to the cancellation of what would have been the first world title fight ever held in Portland and the Pacific Northwest – and the golden opportunity of Firpo's career slipped away, never to return.

Many years later, Firpo's trainer, Mel Epstein, lamenting more with each passing year the significance of the dream of a lifetime slipping suddenly and irrevocably away, would, with emotion, simply say, "Firpo in those years was unbeatable. He was practically unbeatable . . . Who's going to beat him?"

In an interview with the *Los Angeles Times*, in 1934, Joe Waterman, the well-known promoter and manager, who had seen Jack Dempsey, Jack Johnson, Sam Langford and other great boxers in action, stated that Young Firpo is the "Greatest thing in boxing shoes I ever saw. He'll murder every light heavyweight from the champion down!"

Late in his life in 1972, Billy Stepp, the prominent Portland sports editor, named Young Firpo the greatest boxer in Pacific Coast and Pacific Northwest boxing history. At the sunset of Firpo's glorious career, Stepp summed up his ring accomplishments in one concise sentence: "A world's champion was Young Firpo, if ever there was one."

World renowned boxing historian, Vince Colitti, a native of Brooklyn and life-long student of boxing history, ranks Young Firpo the seventh best light heavyweight of the Golden Age of Boxing (the 1930's, 40's and 50's) and fourteenth on his list of the greatest light heavyweights in boxing history. He also ranks Young Firpo second, behind Joey Maxim (born Giuseppe Berardinelli), on his list of the all-time great Italian light heavyweights; and lists him among the uncrowned World Champions of boxing history. Colitti names Young Firpo the greatest fighter that ever came out of the state of Idaho.

In 1974, Firpo was inducted into the Inland Empire Sports Hall of Fame and the Idaho Sports Hall of Fame. In 1975, he was inducted into the New Jersey Boxing Hall of Fame as an honorary member. On November 15, 2008, in Los Angeles, California, Young Firpo, the Wild Bull of Burke, was inducted into the World Boxing Hall of Fame.

Firpo maintained that he engaged in 134 fights, scored 79 knockouts, lost 15, and had four draws.

In 1934, Firpo married Mary Widitz, of Roundup, Montana, and they raised their three children – Cleo Marie Bardelli, Frederick K. Bardelli, and John A. Bardelli – at their home in Osburn, Idaho.

Following his retirement, in 1937, Firpo continued to use his ingenuity, and marvelous strength and endurance, to perform herculean feats of labor for mining companies, including the Hecla and Day Mines, as well as for residents and businesses throughout the Coeur d'Alenes; all of which he accomplished without machinery, using hand tools only – with the indomitable courage and determination he displayed in the prize ring, and the belief that he was working for God.

The Eagles' building in Wallace, Idaho, built in 1905. Guido Bardelli (Young Firpo) fought 12 fights here. He also fought 12 fights at Howarth Hall and one in the ball park. *(Courtesy Butch Jacobson collection)*

The Bardelli family. From left: Fred, became an artist and teacher; Guido, the Pride of the Coeur d'Alenes; Cleo, became a school teacher; Mary, home maker, seamstress and gardener; and John, became an attorney. *(Courtesy Bardelli collection)*

The Wallace Silver Jubilee. During August 1957, the Wallace Silversmiths, of Wallingford, Connecticut, began making plans for their 150th anniversary. They sent out questionnaires to the Chambers of Commerce in all towns of the United States bearing the name of Wallace. There were 44 such states, with two states having two towns named Wallace. In reply, the Wallace, Idaho chamber told them about this mining district. They were impressed to find that Wallace, Idaho is the silver capital of the world and would be celebrating its 75th anniversary, during June 1958. Fred Levering was asked by the chamber to meet with the Silversmiths in New York City, where they agreed to a joint celebration, in which the Silversmiths were willing to spend $40,000 in national advertising through magazines, radio, and television. As a result, 50 newspaper, television, radio, and magazine representatives came to Wallace and spent several days enjoying Jubilee activities, while being briefed on the area's mining industry, its past successes, and current problems.

The Wallace Silver Jubilee float, depicting an early day mining scene where the ore car is emerging from the Free Coinage Mine. Seated on the car is Queen, Cleo Bardelli (Clizer), and standing, Princesses, Barbara Britt (Areitio), and Camille Betts (Oliver). The original design of the float was developed by Ray Giles, employee of the Day Mines, Incorporated, who also built the paper mache donkey. The float was constructed under the supervision of L. A. James, with the assistance of employees of the Coeur d'Alene Hardware and Foundry Company. The float appeared in the Spokane Lilac Festival Parade, on May 17, 1958 and in the Wallace Silver Jubilee parade, on June 14. *(Courtesy Clizer collection)*

Silver Jubilee Queen, Cleo Bardelli, enjoys the first dance with Idaho State Governor, Robert C. Smylie, following her coronation by the governor at the Wallace Civic Auditorium. *(Clizer collection)*

Wallace High School graduates, Cleo Bardelli and Gary Clizer, May 19, 1955. They married June 27, 1959. *(Clizer collection)*

Silver Jubilee Queen, Cleo Bardelli, wields a giant silver knife to cut a mammoth cake, on June 13, 1958. A newspaper article, titled "1000 pound Jubilee Cake to be cut," stated the following:

The multi layered Silver Jubilee birthday cake, which will be cut at 5:30 tomorrow at Sather Field during the barbecue, is 6 feet across and 8 feet high. Along the sides, mural – like panels will depict the history of Idaho's Silver Mining industry. Satin icing, "sparkle" sugar, silver dragees and leaves will impart the "silver" motif. The Syringa, Idaho's state flower, will be employed as a design, and there will be many actual blossoms on the cake. The frosting will be a silvery white.

The cake weighs about 1000 pounds. A total of 625 pounds of ingredients went into the cake: 165 pounds of cake flour, 6 pounds of salt, 10 pounds of baking powder, 200 pounds of sugar, 46 pounds of shortening, 128 pounds of milk and 71 pounds of egg whites. Rice's Bakery personnel made the cake.

A flat-bed truck will transport the decorated cake from Wallace to the scene of the cutting ceremonies. Marie Wood and Carolyn Palmer, Homemakers of Tomorrow for Idaho in 1957 and in 1958, will make the presentation of the cake. Assistant Secretary of the Department of the Interior, the Honorable Royce Hardy, will also take part in the cake cutting ceremonies. The largest silver knife in the world has been cast by Wallace Silversmiths for this occasion. *(Courtesy Cleo Clizer)*

Miss Landsburrough's fourth grade class in Wallace, 1946-47. Back row from left: Jack Sorenson, unknown, Vance Ross, Dave Pasold, Fred Uhl, Bob Perkins, Merlyn Clark and Allan Ruffier. Middle row: Priscilla Loomis, Kaarlene Anderson, Kay Kimbell, Juanita High, Margy Marra, Jean Cleveland, and Fay Spalding. Front row: Kay Nelson, Johnny Drew, Gary Clizer , Alice Giroux, and Ardith Shanks. *(Clizer collection)*

Coach Harold Goldstein's, "under 100 pound" junior high school basketball team, in Wallace, 1950-51. The team was undefeated in both league and tournament play. Back row from left: Jerry Joe Caron, Don Preston, Jim Ostrander, and Wayne Tomlin. Front row from left: Spec Fanning, Bob Yost, Jim Poindexter, Gary Clizer, Gary Trenary, Coach Harold Goldstein. (Missing from the photo is Bud Grant.) Lower right: Jeff Clizer, Gary and Cleo's son, picking huckleberries.

Brothers – Dr. Ed Clizer and pharmacist Gary Clizer. *(Clizer collection)*

Helen and Ernie Clizer, on King Street in Wallace. *(Clizer collection)*

Ernie Clizer, Clizer's Wallace Florist, Cedar Street, Wallace, April 1928. *(Clizer collection)*

A Brief History of the Wallace Mining Museum

Harry Magnuson grew up in Wallace and over the years developed an ever increasing interest in the history of the region. He also was a stockholder in a number of mining companies.

Fred Levering came to Wallace from Olney, Illinois, and also became enamored with the history of his new surroundings. He won a seat on the Wallace City Council and gradually increased his involvement in local and regional activities. He became a novelty of sorts for another reason. He looked just like Harry Truman!

In 1956, Harry came up with the idea of partitioning off the front of the old Silver Dollar building at 507 Bank Street. The back of the building would continue to provide storage space for the firm's mining records, but the front portion was reproposed as space for a museum. Fred Levering was the first director. It wasn't long before enough artifacts became available that the limited space in that tiny "storefront" became an issue. When the Yellowstone Trail garage near the east end of Bank Street closed in the 1960s, the museum moved into the much larger showroom portion of the building.

A couple more changes in the late 1960s and early 1970s resulted in an opportunity that put the museum on the path towards self-sufficiency. Rice's Bakery closed its business, and the location at 509 Bank Street was far better for tourist activity than the other end of town. Besides, the county was buying the old garage as a potential jail and safety facility. Its location across the street from the courthouse made that spot nearly ideal.

Funds were raised with the idea of purchasing the old bakery building and create a permanent home

The Wallace Mining Museum at 509 Bank Street, open five days a week from 10:00 a.m.–3:00 p.m. *(Bamonte photo)*

Fred Levering, first director.
(Courtesy John Amonson)

for the museum. There was plenty of room for exhibits and storage, and I-90 traffic still had to go through town and right by that location.

By this time, a new executive director was managing the facility-a former native of Burke who had grown up in that rough and tumble mining town and had the conversational skills to relate those stories to the visiting public! Her name was Marie Driscoll. Also during this period, the location served as the office for the Wallace Chamber Of Commerce.

The chamber board of directors eyed the gradual increase in tourist activity with considerable interest. One major element of its economic development strategy was to increase the number of tourist dollars flowing into the local economy. The cyclical nature of metal prices and idle mines were not conducive to a robust business model, and much effort was put forth to develop alternatives.

The museum became increasingly important to the community as this tourist-based activity increased. When the Sierra Silver Mine Tour was created, the museum became a logical gathering place for those wishing to see what underground mining was all about. Even when the "storefront" of the Silver Dollar Mining Company building (the first location of the mining museum) became the ticket office for the mine tour, the museum still made its rest rooms available, as well as a warm lobby to keep customers comfortable on colder days.

When Jack Hull was conducting his "gold fields" tour, the museum became the staging area for that venture as well.

Wava Beehner was Marie Driscoll's assistant during many of those transitional years, and when it was time for the aging but still young-at-heart Marie to step down, Wava stepped up into the dual position of Chamber Of Commerce office manager and executive director of the museum.

During this period, however, revenues were not keeping up with escalating expenses, and the museum's board of directors had to face some stark realities. Raising admission fees enough to balance the budget could easily be counterproductive, driving some potential visitors away.

The arrangement with the chamber also did not appear to offer much flexibility. It covered 40% of the executive director's salary, but did not contribute anything for employees needed during the summer.

Marie Driscoll, second director. *(Courtesy John Amonson)*

Wava Beehner, third director. *(Courtesy John Amonson)*

The directors gave serious thought to closing the doors and letting the chamber solve its half of the problem on its own. One of those directors, however, thought that the situation could be salvaged. Mayor Maurice "Mo" Pellissier was on that board, and also served on the board of the Sierra Silver Mine Tour. He was aware of the interest the former mine superintendent, John , had in local history, and with his extensive background in mining practices and techniques "Mo" believed those skills could be instrumental in turning the situation with the museum around. He called John and encouraged him to submit his resume' when the ad for the museum director position was published in the local paper.

John was the successful candidate. He started work February 6, 1990. In his own words he said it was "like learning to swim by being pushed off the dock"! He was quick to clarify that as not stated as a complaint, but the intensity of learning so much in such a short time was quite a challenge!

In addition to learning the ropes as the chamber's office manager and executive director of the museum, he was placed on the committee to prepare for Idaho's statehood centennial. Harry Magnuson was the chairman of the state's centennial commission, and was also the chairman of the mining museum board of directors.

In addition to his already full plate, John was charged with making the museum financially viable. He was aware that most museums and non-profit attractions had gift shops. He was also aware that the law required that 501 (c)(3) non-profits not use that status to undercut the pricing strategies of for-profit competitors. John's own plan was to actually charge higher prices for his merchandise. "Most people will pay $10.00 for an item that is at their fingertips over walking two blocks to get the same thing for $9.50" he said. The scheme worked perfectly! By the end of the first year he was no longer operating at a loss, and that continued that way for the rest of his 17 ½ years in that position.

John's starting salary in 1990 was $1,000 a month. As low as that was it was still a livable sum. In fact, when it came time for a raise, he was less than enthused! "The reality is, that every time the board raises my salary, I have to work harder to make

John with his wife, Linda. John was the fourth director. *(Bamonte photo)*

Jim McReynolds, fifth director. *(Courtesy John Amonson)*

enough money to balance the books" he was quoted as saying. That's not much of an incentive!

John used to claim that he liked to look at the big picture. He liked his weekends off, but also tried to make his employees feel comfortable working there. A good example is Mother's Day. During his early years in that position, he had ladies for his assistant managers, and thought it proper to schedule himself to work Mothers Day.

As luck would have it, a tour bus was coming through Wallace looking for a "pit stop". It was just after 8:00 AM, John's normal summer, opening time. The driver noticed the sign stating "free public rest rooms." That time on a Sunday, parking a full size bus was not a problem. Anticipating some high-end gift shop revenue, he quickly replaced his admission sign with "free admission to tour bus customers". The snowball effect of that impulse was beyond even his lofty expectations.

It turns out that this bus was one of many in a fleet that transported customers from Salt Lake City to Vancouver B.C., to take a cruise ship to Alaska. Those same coaches brought similar groups back to Salt Lake after their own cruises.

This driver shared his experience with other drivers, and the consensus was that Wallace was a perfect distance from their overnight stay and early departure from Missoula. All drivers adopted the pattern. In a relatively short time, the fleet adopted a stop in Wallace on the return trip as well.

While few Wallace merchants were open that early on the westbound stops, the eastbound business was lucrative! The tourist dollars provided a significant boost to a struggling economy!

John also brought some video production experience to the job as well. He created a 20 minute video on the history of the mining district titled "North Idaho's Silver Legacy." He upgraded the projection equipment in the video room, added an 8' wide screen, and included the show in the admission price. He also made free copies of it available to any tour bus operator who wanted one! The result of this effort gave tour groups from all over the world a documentary on the Silver Valley's rich history!

John also went to great additional lengths to educate the public on the region's history and mining heritage through meaningful exhibits and interpretive language upgrades. The president, Don Springer, a local geologist had made good progress in the initial attempt to improve the signage, but John could see that much more progress could be made.

"When you frequently get the same questions on a piece of equipment or practice, you know that the public is best served when you could answer them in advance with concise written explanations" he used to say. "Besides, not all of our employees have the background to give a meaningful response".

Although John was comfortable using state-of-the-art video editing equipment, he was not interested in computers. He once made the comment that he would choose to have his teeth drilled on without anesthetic than to have to stare at a computer screen for any length of time. He did a handful of projects with the computer, but thought it was vastly inferior to his manual system. For example if someone came in on Labor Day and inquired how much taxable sales he had on July 4, he could have an answer in less than ten seconds. Most computer jockeys couldn't match that fast of a response.

As the years unfolded, John began to plan for retirement. Although he was pleased with his work and his accomplishments, he knew there would come a time when he would turn the museum management over to someone else. He was aware that he would qualify for Social Security at age 62, and decided that his exit would coincide with that milestone. By then, he would have had 17 ½ years at the helm, which remains the longest period of any of the past executive directors. In his own words he summed it up this way " I look at myself as a track and field relay runner in the middle of the order. You grab the baton and run with it to the best of your ability, then hand it over to someone else to finish the race".

Well in advance of John's retirement date, the directors advertised for a qualified replacement. A relatively small number of resume's were received. One candidate had underground experience and was a Silver Valley native. Within a short time Jim McReynolds became the successful applicant. Jim was retired from the military, but wanted to stay active. Unlike John, he had a keen interest in computers and convinced the board that digitizing the collection was the logical way to manage the information.

Jim remained at the helm for eight years. During his leadership, he was recognized for his various accomplishments with a number of awards. His management of the collection and archives was considerably more refined and professional than that of the more primitive system used by John . Future researchers will find his system to be more user friendly that of his predecessor.

However, Jim also had other plans for enjoying his retirement, and a planned move to Portland, Oregon would mean turning over the leadership at the museum to its current executive director, Tammy Copeland.

While Tammy did not have a background in hard rock mining, she had grown up in the Kentucky-West Virginia coal belt and had some understanding of the personal side of what it is like to be a miner. Her computer and management skills were adequate for the job, and her enthusiasm was apparent. With Liz Stepro as her assistant, the museum continues to prosper and move forward in this rapidly evolving technological world. There is every indication that the organization will continue to be a productive cornerstone in Wallace's economic future.

Tammy Copeland, sixth director. *(Courtesy Tammy Copeland)*

The Brothels Made Wallace Famous

From the beginnings, the future town of Wallace was, as Colonel Wallace had predicted, an excellent geographical location in the heart of the rich silver mining district. This area included an amalgamation of mining, logging and all other business, all relating to needs for service. It was also an excellent, and central location within Shoshone County to conduct County business. For that same reason it was also chosen as the County seat.

As a result of this, especially the mining activity, young men were coming to the Coeur d'Alenes en masse to work in, mostly the many mines, but also other industries springing up around the area. From 1884 into the 1980s, when the majority of the mines were all running with full crews, the population during those times largely consisted of more males than females. Due to the common characteristics of the people drawn to this area, a large majority being young, single, and high spirited, prostitution became one of the central drawing cards for that geographic region, and stayed that way for many years to come. Consequently, in time, Wallace, Idaho, would become famous for both being the Shoshone County Seat, and its phenomenal brothels. **The following mine names illustrate the amount of activity in the Coeur d'Alenes which began in 1883.**

The Coeur d'Alene Lead-Silver Mines

A List of the Principal Mining Claims in the Coeur d'Alenes from 1884 to 1900

The lead-silver mines of the Coeur d' Alenes make up three distinct geographical districts along the valley, the western, central, and eastern sections.

(Taken from Transactions of the 1903 ***American Institute of Mining Engineers, Volume XXXIII****)*

1. Manila
2. Sunny
3. O. K. Western
4. Whippoorwill
5. Mabel
6. O. K.
7. Johnny
8. Danish
9. Reeves
10. Packard
11. Royal Knight
12. Queen
13. Eureka
14. Crown Point
15. Henry
16. Silver King
17. Senator Stewart
18. Quaker
19. Silver Casket
20. California
21. Idaho
22. Legal Tender
23. Harrison
24. Princess
25. Sugar
26. Fifty-Three
27. Florence
28. No. Five
29. No. One
30. No. Two
31. No. Three
32. No. Four
33. Ontario
34. Ontario Fractiou
35. Carbonate
36. Sierra Nevada
37. Tip Top
38. Apex
39. Rambler
40. Oakland
41. Excelsior
42. Cariboo
43. Good Luck
44. Butte
45. Hard-Cash
46. Hamilton Fraction
47. Hamilton
48. Nevada
49. Link
50. Arizona
51. Viola
52. San Carlos
53. Likely
54. Skookum
55. Jersey Fraction
56. McLellan
57. Lilly May
58. Deadwood
59. Sought Again
60. Coxey
61. Debs
62. Carter
63. Josie
64. Allie
65. Jack Ass
66. Sold Again
67. Kirby
68. Shoshone
69. Summit
70. Miles
71. Teddy
72. Republican
73. Tyler
74. King
75. Pitt
76. Cheyenne
77. Wasp
78. Wheel Barrow
79. New Era
80. Bee
81. Hornet
82. Hawk
83. Caledonia
84. O'Connor
85. Omaha
86. Scorpion
87. Oregon
88. Combination
89. Holman
90. Butler
91. Kaintuck
92. Dakota
93. Capetown
94. Emma & Last Chance
95. Puritan
96. Idaho
97. Utah
98. Last Chance
99. Sampson
100. Helen Mar
101. Stem Winder
102. Emma
103. Kruger
104. Cuban Mines
105. Silver
106. Johannesburg
107. Forest Belle
108. Bunker Hill Mines
109. Bunker Hill
110. Important
111. Phil. Sheridan
112. Chestnut
113. Ivanhoe Fraction
114. Blue Bird
115. Geyser
116. Susie
117. Maple
118. Rookery
119. Homestake
120. Butternut
121. African Reef
122. Mary
123. Beulah
124. Zululand
125. Matabela-laud
126. Stopping
127. Buckeye
128. Reed
129. Sullivan
130. Small Hopes
131. Lackawanna
132. Alla
133. Mashonaland.
134. Mabundaland
135. No Name
136. Miners Delight
137. East
138. Sullivan Fraction
139. Rolling Stone
140. Daisy
141. Fair View
142. Schofield
143. Ollie McMil
144. Bonanza
145. Iron Hill
146. La Crosse
147. Summit
148. Justice
149. Jessie
150. Comstock
151. Ophir
152. Dandy
153. Julia
154. Walla Walla
155. Midday
156. No. 1
157. Lucky Chance
158. Bonnie Jean
159. Tough Nut
160. Iuka
161. Milo
162. Fannie May
163. Protection
164. Alhambra
165. Dawn
166. McKinley
167. Monte Cristo
168. Crescent
170. Southern Cross
171. Omega
172. Polaris
173. Step-and-a-Half
174. Chester
175. Protection
176. Hanna
177. McKinley
178. Bartlett
179. Emma Nevada
180. Coeur d'Alene Nellie
181. Argentine
182. Argentine Fraction
183. Keystone
184. Lee
185. Brewery Mines
186. Brewery
187. Electric Light Co.'s Gr
188. Big Medicine
189. Union Mines
190. Vermont
191. Oreornogo Mines
192. Blue Tauck
193. Brother
194. Sister.
195. Oro Fino
196. Silver Wave
197. Sunshine Mines
198. Black Cloud Mines
199. Sunshine
200. Iowa
201. Monarch
202. Black Cloud
203. Pacific
204. Ninety-Nine
205. Contact
206. Mountain Chief
207. Shamrock
208. Black Prince
209. Divide
210. Marietta
211. Hornet
212. Idaho
213. Panhandle
214. Ohio
215. Bonanza
216. Nugget Placer
217. Johnny Placer
218. Figaro
219. Banner
220. Lucky Boy
221. Iron Mask
222. Idaho Mint
223. Lucky Boy Fraction
224. Yellow Jacket
225. Free Coinage
226. Speckled Trout
227. Anna
228. Topeka
229. Pierce
230. Mell
231. Custer Mines
232. Star
233. Treasure Vault
234. Mingo Chief
235. Kola
236. Belle of the West
237. Tugela
238. Belle Fraction
239. Trade Dollar
240. Granite
241. Denver
242. Crystal
243. Empire State
244. Manila
245. Manila Fraction
246. Old Hickory
247. Daisy
248. Patuxent
249. American Eagle
250. Eagle Fraction
251. Bi-Metallic
252. 16-to-1
253. Hawaiian
254. Ambitious
255. Lucky Jim
256. Yew
257. Miller
258. Aspen
259. Jenkins
260. Chilcot
261. Rosa
262. Rosa Fraction
263. Mace
264. Mascot
265. Blarney Stone
266. Knapp
267. May Flower
268. Protection
269. Dobson Jim
270. White Quartz
271. Estelle

272. Virginia Placer
273. Idaho
274. North Star Mines
275. Grace
276. Minneapolis
277. Sunlight Placer
278. Fidelity
279. Coeur d'Alene Chief
280. Morning Light
281. Emma A.
282. Fanny Peak
283. Faithful
284. Hopeful
285. Security
286. Dorothy Hill
287. Carlisle
288. President
289. Lucy
290. Merrimac
291. Monitor
292. Monitor Fraction
293. Manhattan Fraction
294. Mountain Goat
295. Amazon
296. New York
297. Staten Island
298. Ajax
299. Father
300. Diamond Fraction
301. Chloride Queen
302. Switch Back
303. Roy
304. Roy Fraction
305. Live Oak
306. Tough Nut
307. Colwyn
308. Tuscumbia
309. West
310. Parallel
311. Sitting Bull
312. Silver Tip Fraction
313. Parrot
314. Mule Deer
315. Silver Tip
316. Red Dragon
317. Sun Set
318. Big Bug
319. Ula
320. Try Me
321. Nilus
322. Leonard
323. Polly
324. Tamarack
325. Tamerlane
326. Chesapeake
327. Carbon
328. Custer
329. Pacific
330. Timber Line
331. Rose
332. Diamond
333. Green Hill
334. Cleveland
335. Green Hill Fraction
336. Grey Eagle
337. Head Light
338. Mammoth
339. Tariff
340. Mammoth Fraction
341. Little Chap
342. Snow Line
343. Saturday
344. Walter McKay
345. Bonaparte
346. Sancho
347. Crown Point
348. Fairview Fraction
349. Sherman
350. Union Fraction
351. Union
352. Bengal Tiger Fraction
353. Hidden Treasure
354. Burke Tiger
355. Stanley
356. Silver Chief No. 2
357. Wide West
358. Shoo Fly
359. Tiger
360. Sheridan
361. Bullion
362. Galena
363. Bullion Mines
364. Liebes Fraction
365. Reeves Mines
366. Diamond Hitch
367. Seattle
368. Wellington
369. Banner Fraction
370. Banner
371. Silver Queen
372. Anchor
373. Youngstown
374. Bonanza
375. Sandwich
376. Sullivan
377. Standard
377a. Combination
378. Gray Copper Fraction
379. Gray Copper
380. Crown Point
381. Tom Reed
382. Selkirk
383. Akron
384. Broad Gauge
385. Narrow Gauge
386. Chief
387. Josie
388. Gem of the Mountains
389. San Francisco
390. Esler
391. Yankee Doodle
392. Esler Fraction
393. Apex
394. Cape Horn
395. Idaho
396. Black Bear
397. Badger
398. Gem Fraction
399. So Mote It Be
400. Moon Light
401. Grover Cleveland
402. Black Bear Fraction
403. Protection
404. Queen of the Hills
405. Grubstake
406. Grubstake Fraction
407. Bell
408. Contention
409. Bell Fraction
410. Climax
411. Iron Silver
412. Josephine
413. Tuff Nut No. 2
414. Tuff Nut
415. Exchequer
416. Mountain Fraction
417. Muulton
418. Mountain View No. 2
419. Mountain View
420. Senator
421. Senator Fraction
422. Parrott Fraction
423. Buffalo Fraction
424. Denver
425. Oreornogo
426. Hecla
427. Sunday
428. Poor Man
429. Burke
430. Ella
431. Fuller
432. Protection
433. Green Mountain.
434. Got-'em-Now
435. Mat
436. Parrott
437. Sonora Mines
438. Modock
439. Sonora Fraction
440. Sonora
441. Alameda
442. First Chance
443. Great Hope
444. First Chance Fraction
445. May Flower
446. Clark
447. O'Neil
447a. Columbia
448. Mond
449. Russell
450. Ironside
451. Consolidated Ext.
452. Katie May
453. Orphan Boy
454. Orphan Girl
455. Muscatine Faction
456. Burlington
457. Cresus
458. Muscatine
459. Star
460. Leadville Fraction
461. Leadville
462. Oreornogo Fraction
463. Dipper
464. San Jose
465. Little Strike
466. Iron Crown
467. Iron Crown Fraction
468. Lauren J.
469. Hickory
470. Grouse
471. Noon Day
472. Morning
473. Silver King
474. Evening
475. Midnight
476. You Like
477. Bummer
478. Silver Queen
479. Park
480. Independence
481. Fanny Gremm
482. Still Water
483. Lucretia
484. Key
485. Contact
486. Gettysburg
487. True Blue
488. Buckeye
489. War Dance
490. Victor
491. Lost Wonder
492. Victor Fraction
493. Jersey Minor
494. Pay Master
495. Ada
496. Yolande
497. Ida
498. Clear Grit
499. America
500. Cuban Republic
501. Mary Norem
502. Commander
503. Gold Hunter
504. Away Up
503. Joe Dandy
506. Enterprise
507. Bullion
508. Pine Tree
509. Defiance
510. Mullan
511. True Blue
512. West Mullan
513. Midnight No. 2
514. Hennessey Mines
515. Ryan Mines
516. Thos. Brenau Mines
517. P. M. Kavaua
518. Central
519. Cape Nome
520. Crown Mines
521. Morning Mines
522. Evening Mines
523. Hattie
524. Klondike
525. Snow Flake
526. Club
527. Gold Queen
528. Hammerfest
529. May Flower
530. Cape Nome
531. Surprise
532. Sunshine
533. Sunshine Fraction
534. Gold Drop

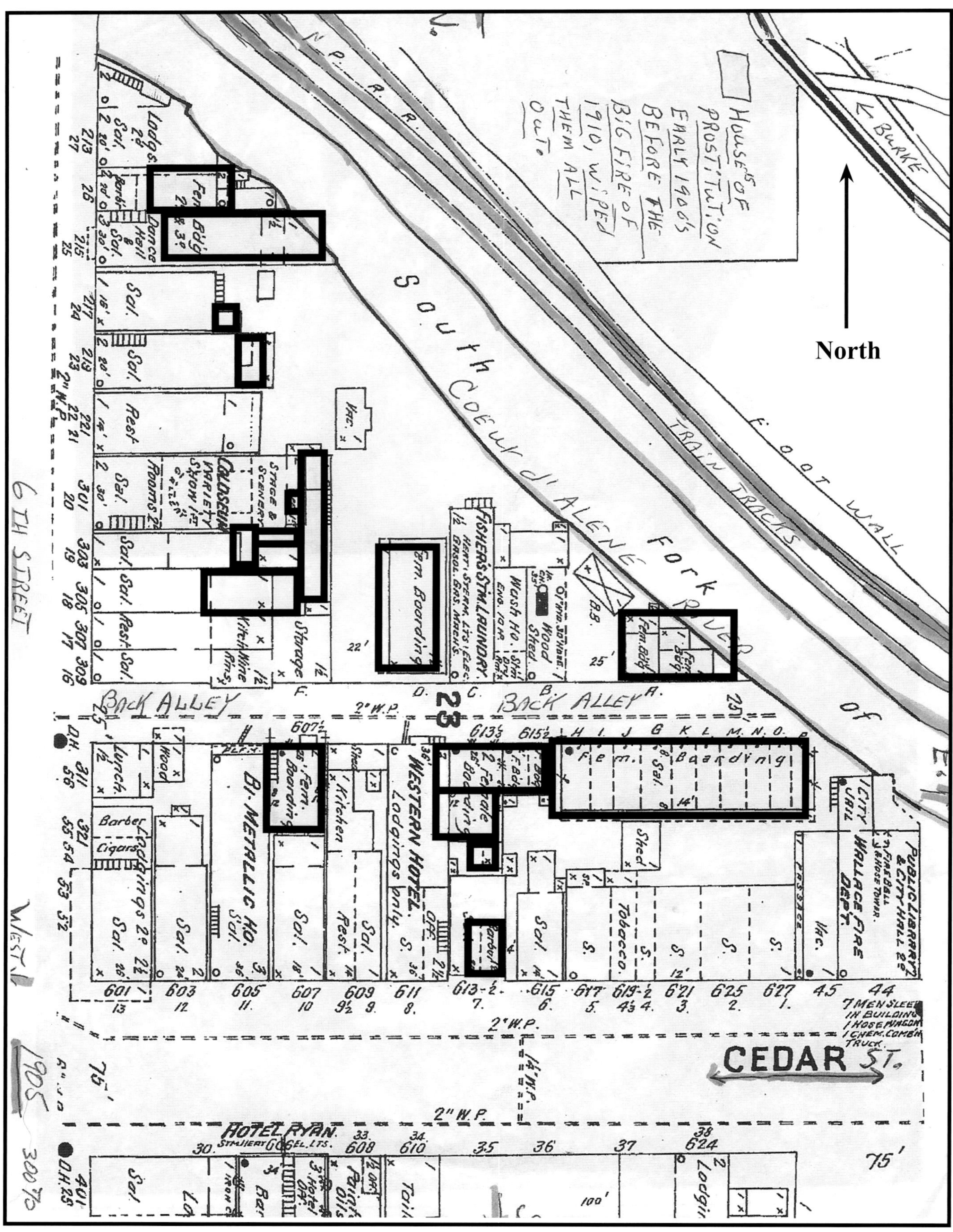

The brothels in Wallace, Idaho in the early 1900s. They are marked with the dark outlines. Original source – Sandborn Fire Map, 1905. *(Diagram courtesy Butch Jacobson)*

Within a 100 year period, up to the 1980s, the Coeur d'Alene Mining District produced a billion ounces of silver, half of the nation's newly mined silver. The Coeur d'Alene Mining District was largely inhabited by what was called "tramp miners," living from paycheck to paycheck and drifting from mine to mine. Their lifestyle typically involved an abundant use of alcohol, visits to brothels, and barroom fights. Life for them was tough and exceptionally dangerous.

In a book written by the late Gene Hyde, *From Hell to Heaven, Death-Related Mining Accidents in North Idaho*, Gene covered every mining death from February 19, 1887, to June 5, 2001. The first recorded fatality, of Lyman Wilson, age 25, in 1887, was by a dynamite explosion at the Hornsilver Mining and Milling Company, three miles up Placer Creek from Wallace. Gene's last entries were the deaths of Wayne Leroy Brenner, 34, and Perry Neil Stack, 47. Both were buried alive as the result of a rock burst in the Galena Mine.

Within the small geographical area of the Coeur d'Alene Mining District from 1887 to 2001, a total of 3,238 miners were killed by mining accidents. It was easy to see why the brothels in Wallace were an integral part of the community and appreciated by many.

In 2017, Dr. Heather Branstetter wrote an excellent, well-written and researched book titled, *Selling Sex in the Silver Valley, A Business Doing Pleasure*. With her permission, I have quoted a number of pages from her book, which I feel best describe how Wallace was able to overcome, for over 105 years, some of the many obstacles involved with prostitution and still stay in business. The following was taken from chapter 4, pages 85 to 94:

"You Don't Have to Obey the Laws, But You Do Have to Follow the Rules"

Almost every person I talked to during my oral history interviews repeated phrases like "The houses offered relief for single miners and kept local women from getting raped," "The women were checked out by the doctors and didn't solicit around town or on the streets" and, especially, "The houses gave so much back to the community." After Dolores [a madam] expanded her operations with the Luxette, people used to say, "It's even become corporate." Pithy puns and jokes consisting only of a single punch line, both innocent and suggestive, were repeated freely in Wallace. For example, many people related versions of a clever story about bar owners beneath the houses complaining, "Business is great, but there's too much fucking overhead." Another man told me about running into the father of his girlfriend at the time as he was coming back down the stairs on his way out the door. His apprehension quickly dissolved when the father said to him, "Boy, am I glad to see you here." Men who were grocery boys back in the days love to tell stories about their jobs delivering to the houses, which were known for handing out generous tips.

It's no accident that such phrases and stories are the chorus around town to this day. The madams appealed to the sense of humor and common sense values of the community, both in language and in deed, as they set the tone to subtly align the city's rhetoric about the houses. As former Day Mines and Sunshine Mines employee Rod Higgins explained, Wallace's embrace of the commercial sex industry was the result of the brothel managers' purposeful persuasive strategy: "Everyone viewed them [the brothels] as being an advantage, thanks to the madams' marketing efforts." Men and women alike repeatedly told me how "fun" Wallace was during this time, as they expressed gratitude for having the opportunity to grow up in such a great town.

Talking points about the benefits of the sex work in Wallace also offered direct rebuttal to threatening critiques of prostitution that arose periodically during the years. Specific practices, such as doctor's visits, solidified the madams' words into a "code of conduct" and built a foundation for the town's present-day understanding of its past. It was said that the brothels: 1) operated in accordance with a "live and let live" sense of order, providing services that helped keep the local streets safe at night; 2) were restricted to the northeast corner of town, where they were regulated and mostly free from sexually transmitted infections; and 3) contributed to the economy in Selling Sex in the Silver Valley both direct and indirect ways. Written records of history affirm variants of these arguments, which persisted through the years. The industry served local miners and regional loggers, contributed to the schools and attracted business from truckers, college students and Canadians, drawing men and their money into the community from distant places.

In Wallace, it was said that prostitution's "image was different—the image they presented and your image of them was different. The community had a high degree of acceptance of the girls and the business." One newspaper article featured a sex worker saying she preferred Wallace to Nevada, where she encountered "a rougher clientele" and was not as "free to come and go as she pleased," unlike in Wallace, where she could "come up [and work] when I feel like it, work till I'm tired of working, and then I go home." By the mid-1950s, the town of three thousand people would elect only city leaders who allowed the prostitution industry to continue, driving away the few pastors or priests who advocated for reform. As late as the 1980s, former mayor Pellissier told me, a group of women he called the "Golden Girls" said they would not support his bid for mayor unless he promised not to shut down the sex industry; they threatened to campaign against him if he closed the houses and they'd convinced him they wielded enough influence to derail his election.

Evidence for the residents' continuing acceptance of the brothels is not limited to quotes appearing in old newspaper articles and the oral histories I conducted. A study conducted by Robert Miles, who looked at prostitution in Wallace for his master's thesis in criminal justice at Washington State University, documents community perception during the late 1970s. Miles's research revealed that both men and women in Wallace accepted the houses as integral to the town's way of life. Throughout the study, a large majority of respondents agreed that prostitution was not morally wrong, did not lower respect for women and should be decriminalized. For example, 69 percent of respondents disagreed with the statement that "decriminalizing prostitution would seriously weaken the social fabric of our community," and 75 percent disagreed with the idea that "decriminalizing prostitution decreases respect for the institution of marriage." In 1984, historians Hart and Nelson wrote that Wallace's "unabashed and unrepentant acceptance of drinking, gambling and whoring, seen as commonplace...even acquired a certain civic respectability." What's more, they added, people in the community believed that "prostitution kept together many respectable marriages, that Wallace madams gave generously to deserving charities, and that revenues from prostitution kept the city coffers filled and the streets paved." The madams framed their business as central to the community and then followed up by donating in visible ways.

The houses had historically been a central part of the economic life of Wallace, but madams increased the publicity and scope of their monetary contributions to the town during the 1940s and 1950s. The women wove their work into the structure of the community by circulating small talk that appealed to values like civic service and philanthropy—appeals that were represented by monetary contributions that supported the town's families, churches and schools. During the 1940s, Magnuson told me, the Salvation Army used to collect every Friday night from the houses and the card tables, and they "did a lot of good with that money." Many research participants told me the madams also contributed in less visible ways, citing as an example the way they would quietly purchase food and leave it at the grocery store available for the clerks to distribute to families in need according to their discretion.

It is difficult to find people who opposed the industry, but Holly Shewmaker, who went to high school in the valley during the early 1960s, spoke with me at length about some of the ways the presence of commercial sex influenced the lives of women who were not madams or whores:

"Sex work affected my marriage and degraded my ex-husband, who taught me things based on what he'd been taught up there. "Being with you was like being with a courtesan," he once told me. I came to think sex was something you did in service of men, got pregnant at sixteen, and had a baby at seventeen."

It would be a mistake to glorify the madams in any way, Shewmaker cautioned me, because their profession was "based on lies. Prostitution is a lie and it erodes the soul. It destroys something meant to unite people." Shewmaker added that her dad had "tried to make me feel better because he was a city councilman. 'Whores built the viaduct,' he said."

Even though most women in town believed the sex industry was in their interest to promote, some women felt a tension under the surface because it "was like a secret club in this town between the men and women in the houses, but the other women in Wallace weren't allowed into the boys' club. Dolores had a relationship with men that other women were excluded from." Visitors from out of town used to drive by and catcall young girls as they walked down the street. Some men from the region presumed most women in town were sex workers and were blatantly disrespectful when a Wallace girl revealed where she was from. Shewmaker related a story about high school cheerleaders participating in a Spokane parade: "Wallace girls marched in their black sweaters with the large "W" [for Wallace] across their chests. Coins were tossed at them and jeers of "W" for whore" rang out.... At this point, I had come to believe I would rather be a madam or a whore than one of the "protected" women of Wallace."

She added that one of the girls went home and recounted the story to her mother, who told her never again to speak a bad word about those women because they protected young women from being raped and supported the community economically, including their own family's business.

These stories demonstrate how negative perceptions were countered through sayings that became folded into collective memory, layered within narratives repeated over time and embodied by the madams' respectability in deed as well as discourse. The women who worked upstairs knew that word would travel as long as they maintained a consistent image. And word did spread. The situation in the Silver Valley became widely known. One of my research participants told a story about working on a radio show in Los Angeles during the early 1970s; when the host, Sammy Jackson, asked her where she was from, she answered, "You would not know where I'm from, Sammy. I mean, it's a small town in northern Idaho." After she finally admitted she was from Wallace, Jackson answered, "Oh baby! I spent the best month of my life in Wallace, Idaho." Then she added to me, "I was just stunned that he would even know anything about northern Idaho."

When reporters from around the region would come to town to do a story about the houses, the madams promoted their business practices openly and in accordance with the town's values. Emphasis on the women's care for single men and physical cleanliness countered the stereotype that sex work is immoral and spreads disease. One Seattle Times newspaper article, for example, quotes a madam saying that the "men need relief" and "are well taken care of here," featuring a photo of a neatly prepared bed and narration promising a "spotlessly clean room" because the madam "will not have anything dirty!" The women created the perception that their profession was advantageous to the town because the industry operated in an orderly fashion according to community wishes. An article appearing in Boise's Idaho States-

man features one of the town's female residents professing, "A mining town needs brothels" and quotes the manager of the U&I Rooms promoting an understanding of justice congruent with the town's historic mining camp roots: "You don't have to obey the laws, but you do have to follow the rules." The women were said to bring money into the area and keep that money local. Most often noted, both in the newspapers and around town, was the way in which the sex industry supported the schools. In 1991, the Seattle Post-Intelligencer reported that "madams, for example, helped buy the scoreboard for the high school football team. They also pitched in to help with youth baseball uniforms." Madams notoriously bought the school's band uniforms, and rumor has it that they funded the uniforms partly in exchange for an agreement that the band would no longer march around the streets to practice early in the morning. The houses were said to provide good jobs for women in the community, including those who worked as maids in the houses and others who supported the sex workers' food, grooming, clothing and transportation needs. One woman told me about how she used to give the women rides back and forth from Spokane to Wallace, where she worked at a bar called Sweets and ran food up to the girls during her shifts. She adds that she also "did make a lot of money selling Avon to the girls. They were great customers." The madams' indirect economic contributions to the town and the residents' positive small talk about these contributions kept the sex industry integrated into the town's social and institutional fabric.

It is possible to estimate the numbers of women who came into town and worked in the houses from 1940 to 1960 or so because Nellie Stockbridge, whose photography studio was just across the street from the U&I Rooms, took pictures of all the sex workers during this time for licensing and record-keeping purposes. These photos are now housed in the Barnard-Stockbridge Collection at the University of Idaho Library Special Collections and Archives. Many are stiff and look like mug shots, but some of them are softer and feature the women in relaxed and smiling poses so they could buy the portraits and give them as gifts to relatives, friends and others. The women were rumored to have been good customers of the Stockbridge business. The Barnard/Stockbridge Collection photos indicate that an average of 30 to 60 women came into town per year. The mid-'50s were particularly busy. During some of these years, 70 or 80 women moved into and out of Wallace's brothels. Old Shoshone County Sheriff's Office (SCSO) records document at least 531 women who worked in the houses between 1952 and 1973, although it's likely that some files were lost over the years.

Policies, Procedures and Practices

The madams operated according to an unwritten set of "rules" that promoted discretion and family stability, and the policies were well known around town. They were also enforced by small talk that became biting and judgmental if community residents began to feel threatened by boundary transgressions. Sex workers were discouraged from wandering around town aside from taking care of personal errands, and they were rather strictly prohibited from hanging out in the local bars. When research participants repeated variants of the rules to me, they were often related along with stories about notable exceptions. "You weren't supposed to date the working girls," but some people did, Penny Michael said, going on to tell a story about breaking into Quinn's Hot Springs with a woman who was dating her friend.

Sex workers also married local men and became homemakers. Although some of these marriages were socially acceptable, others were affected by stigma that haunted the women in gossip, and one family chose to move away rather than deal with the town's talk. Local girls weren't supposed to work in the houses, and if they were discovered to be from the area, they were kicked out. This rule appears to have been enforced rather strictly—the Shoshone County Sheriff's

Office files confirm that any time a local woman tried to get hired in the houses, they reported it to the madams and made a note in the file that she was not allowed to work there. One man believes a girl from nearby Mullan worked in one of the houses, but he's "not sure how she got away with it." This woman may have been the same person described by Tammy Polla, who remembered a girl who came to Wallace to work in the houses and went to Mullan to try to get a high school diploma, which she'd never received. The girl was shamed out of school by the other kids, who made her life miserable after they found out.

Former maid Kristi Gnaedinger described her perception of "the rules" concerning the women's social life:

"They were very discreet around town. They stuck to the rules, but not towards the very end. There was one girl who would come and party with us at my house when she was off. She came over and helped me move to Seattle. They occasionally had boyfriends in the community. They had favorite guys, but if they came up they had to pay. When you were there, you were working. The guys would sit in the rooms, the parlors up front where they could visit and socialize, and sometimes the guys would just come up and drink and drink and hardly ever go back into the rooms. So the girls could socialize, but if there were people who came in and wanted them to work, they would work. It was all money. It was very matter of fact. It was a business."

The sex workers were commonly referred to as "girls," even though they were officially required to be over the age of twenty-one. And the term "maid" in Wallace did not simply refer to someone who cleaned the houses. They were also called housekeepers, in a way that was very literal, insofar as they kept the house in order; they did sweep, mop and vacuum, but they also participated in much of the brothel management, answering the door, ensuring everyone stayed busy and knocking on doors to notify the couples that time was up after the timers went off. Sex workers set these timers as they dropped off their money into lockboxes after negotiating services and settling on an appropriate length of time for the encounter. The maids also shopped for food and cooked meals. Former maid Dee Greer said she liked to "spoil them a little bit" by taking them coffee when she woke the women up before their shifts.

The maids were primarily women who were from the area, but some local women were even limited from taking on this role as well. Gnaedinger thought her father "put some pressure on them [the city leaders and brothel manager]" to encourage her to quit because he was not happy with "the fact that his daughter was a maid up at the whorehouse." He would later become one of the doctors who conducted the sex workers' health checkups. Gnaedinger explained that she still maintained relationships with the women after she stopped working as a maid:

"I still went up there to visit. I took my [infant] daughter up there to visit. At the time, Doc Peterson was still the doctor that took care of them and when he retired my dad took over. He really liked Tanya. He was very fond of her. She was a pretty special person. But my mother, probably more so than my dad, was worried about what people would say. "You can't do that. We have a reputation." Get a life, Mom. People were accepting of the houses, but you just didn't talk about it. My mother was worried about the family reputation. Lee told me that the city told her they had to get rid of me."

It wasn't just families worried about reputations who were concerned about the women who were employed as maids. The madams were also careful not to hire someone who could be perceived as personally connected to too many people in the town.

Sue Hansen related a story about applying for a housekeeping job at the Lux when Gennie Freeman was retiring:
"So I went up to see Dolores, and Gennie put me in this room. I remember there was a jukebox in there. It was a nice room. And I waited for a few

minutes and pretty soon, Dolores came in. Beautiful, beautiful, classy lady. And so she shook my hand and we started talking and she asked me if I'd ever done that kind of work before. And I said, well, yes. Because I used to clean for Millie Mara and a few other people. And she said, "Do you have any children?" And I said yeah, I have a little girl, about six months old. And she said, "Are you married?" And I said yes, well, and she said, "Does your husband mind, does he have a problem with it?" And I said, "Oh, no." So about that time Gennie knocks on the door, and she comes in and whispers something to Dolores, and Dolores goes, "Oh...I'm sorry honey, I was interviewing you for a different job."

The maids hustled the clients into and out of the rooms, calling out "man on the floor" when they walked someone from the parlor to the rooms or from the rooms to the door. This procedure prevented the men from running into someone they hadn't come up with. Housekeepers constantly kept tabs on the safety of the girls, and they monitored the patrons closely to prevent harassment or abuse. Greer explained: "At the Oasis they had peek holes in the doors. You tell people that [and they say], "Oh I'll bet you had fun." Well, it was if you heard something, a noise like somebody's trying to rough somebody up. Then you could go peek in the door and see if they were okay. Every place should have had it like that. And in the parlors... you know when you open the door downstairs a bell rings upstairs. And there's a window where you can see down, but they can't see you. And you always looked for somebody who had a sack, checked that out to make sure there wasn't any gun or whatever."

The maids needed to have tough personalities to deal with the customers, who could be disrespectful if they had been drinking too much. Greer told me she caught one young drunk guy pissing on the floor of the parlor room when the door was closed. She kept her cool and grabbed a wash bucket and soap, opened up the parlor door and handed the bucket to the kid, saying, "Now here, scrub that carpet." The guy pretended to be confused and then denied that he'd peed on the floor, to which Greer responded, "I stood here and watched you through the peek hole, dummy. Now—on your knees!"

Both Greer and Gnaedinger had stories about men who tried to talk them into sleeping with them while they were maids. Gnaedinger said when she was pregnant she was uncommonly desirable to some of the clients, especially one who liked "really fat women" and another who wanted her to take him back to her place and tie him up. Greer admitted that the men were persistent sometimes, and she told a story about one guy who had come up with four of his friends:

This guy says, "And I'll have you." I said, "Oh, mercy, honey, if you'd seen my husband you wouldn't want me." And he says, "I don't care. I want you." I said, "Oh come on, you can shop [meaning, find another girl you'd like to be with]."

So they all come in, and he says, "No, I don't want anybody. You can bring everybody in the whole town," he said. And pretty soon the other four took somebody and the madam comes out and we're playing music and I'm fixing him a drink. And he says, "I want you." I says, "You can't have me." And he got clear up to $500 for a quickie, and Billie's going, "Diane, Diane [take the money and do it]." Oh no, oh no, oh no. No. My husband is probably standing right around the corner. . . .

The above quote from Dr. Heather's Branstetter's book titled, Selling Sex in the Silver Valley, A business doing Pleasure, is an unwritten reason why the prostitutes of Wallace were expected to follow the policies, procedures and practices within the community for their own safety (see following story).

An example of when the owner of the U & I rooms, didn't follow the rules

Chapter IX

The Wallace Brothels, Their Sometimes Dangerous Customers, and Other Connections

From the beginnings of Wallace there was a great demand for prostitution – specifically because of the ratio of men to women in the area. As a result the prostitutes in the Wallace area became an integral and accepted part of the community where they lived.

Which brings me to my first personal experience with prostitutes as related to police work. For the first two years I was on the Spokane Police Department, I was often assigned to work the paddy wagon or walk a beat in the downtown area.

When I worked the paddy wagon it looked like an old Wonder Bread truck, the type used during the 50s to 70s. It was a marked unit, painted black and white to match the prowl cars during that time. In Spokane, the paddy wagon was referred to as car #80. The reason it looked like a bread truck was its size and shape. The front, where the driver and passenger sat, was partitioned with solid steel and a small covered peep-hole allowed a view of the passengers in the back, and small enough to keep from being disturbed by the prisoners. This was a two-man unit and always worked the downtown area. It was often referred to as the drunk wagon. The back of this vehicle had two facing bench seats, and another bench-type seat against the steel separating partition. All in all, the paddy wagon had a comfortable seating capacity for approximately 10 drunken and disorderly folks, and possibly 15 smaller, more behaved, people.

Working the paddy wagon was a unique and interesting assignment. Anyone that ever worked it for any length of time would have found it to be a most interesting experience. It gave you the opportunity to meet some of the city's most remarkable down-and-out people and also some of the worst society had to offer. Mostly, it consisted of picking up drunks or those who considered themselves as "tough guys" and typically would resist being arrested.

There are no photos of a Spokane police paddy wagon, but this is a sample. *(Public domain)*

Often, during slow nights we would park in various parking lots, observing any action that may require our attention. It was during the slow times that we would often receive visits from some of the areas prostitutes. They would often stand next to the paddy wagon doorway and talk about problem customers they would sometimes encounter. More than causing trouble or making a point to arrest them, we typically looked out for their safety and they realized that. They were often the victims of perverts who made it a point to prey on them. Most of these women had worked in numerous other places, many in Wallace where they knew they would not get arrested.

Dangerous Work in the World of Prostitution "You Don't Have to Obey the Laws, But You Do Have to Follow the Rules"

In 1971, Butch Jacobson, the person most responsible for my researching and writing this book, and who was a former assistant police chief in Wallace, had an encounter with a man from Spokane who had been pimping some of his better looking girls at the U & I, which stands for You and I, brothel in Wallace. This was a violation of the unwritten but well-understood "rules."

Bringing someone in from out of town to help business in your brothel was taking an unwarranted chance. The person the madam allowed to brought in proved to be exceptionally dangerous and a cold-blooded killer.

By the time I found this out, I had known Butch for almost 20 years and learned of this situation by accident. Inadvertently, when I found out that his incident related to the brothels in Wallace, I asked him if he would write a short story regarding this circumstance. Less than a week after Butch stopped of this man, he committed an armed robbery of the Bon Marche Department Store in Spokane. During the process of the robbery, he shot and killed the head cashier and was in the process of escaping. During his attempted escape he also tried to shoot me.

"A Pimp On His Way to Prison Makes a Stop at Wallace, Idaho"

a story written by
Butch Jacobson for *Nostalgia Magazine*

On October 10, 1971, my partner, Doug Stanley, and I were at the Wallace Police Station, located on the east end of Cedar and 7th street. While walking out of the station, towards our patrol car, we spotted a strange looking man about 70 to 75 yards away. The time was half past midnight, and it was very dark out with the exception of some light from a street light and a lit up business sign.

The man was about six feet tall, slender build, wearing a full length, bright red coat. He had a purple pointed style hat with a very long feather sticking from the band. We walked down the sidewalk and asked this man what he was up to. He stated he was waiting for a friend. He pointed up to the U & I rooms (which is a house or rooms that board young ladies). We knew this man had to be a pimp by his looks.

At this point, I asked the man for some identification, and I remember him saying, "Why are you harassing me? I haven't done anything except stand by my car."

I asked one more time to see some identification, and he handed me his driver's license. My partner called from his portable radio to the Sheriff's Office dispatcher. He requested information on an African American named Jerry Lewis from Spokane, Washington, including his date of birth, and since we

Butch Jacobson, assistant police chief of Wallace, Idaho in 1975. *(Courtesy Butch Jacobson)*

Tony Bamonte in 1972. Jacobson just learned about my connection with Lewis in 2016 through a *Nostalgia Magazine* article. *(Bamonte photo)*

were standing by an early model, black four-door Cadillac, we also gave the dispatcher the license plate number.

A few minutes later the dispatcher called us back, stating that Jerry Lewis had a long criminal record. The car was registered to his mother. At this time, I told Mr. Lewis he had ten minutes to get his friend and leave Wallace, or I was going to arrest him on vagrancy or public nuisance charges.

We then walked back to our patrol car, and drove around a couple city blocks. We hid behind the Hecla Mining office, out of sight, around 125 yards from where Lewis was last standing. Seven or eight minutes went by when we observed a beautiful, young blond woman coming out of the U & I rooms. She had a couple of small suitcases and a handful of clothing. Lewis, after looking around, grabbed the woman's suitcases and threw them into the trunk. The woman threw her clothing and other small items into the back seat. They both got into the car and drove off, heading west out of Wallace.

Throughout all of this we had been keeping dispatch informed of what was going on. We were told not to stop Lewis for any reason, because the FBI wanted to catch Lewis coming over the Idaho/ Washington State line. They wanted to arrest Lewis on federal charges. Other agencies were told to stay out of sight, but to call in when Lewis's car drove past their towns. My partner and I followed slowly behind Lewis, and we lost sight of him when we stopped the car to let pedestrians cross the street.

Ten minutes later an Osburn Police officer called the dispatcher, stating the black Cadillac had not passed by his area.

My partner said, "I believe Lewis smelled a rat trap for him. He has to be hiding somewhere between Wallace and Osburn."

About a quarter of a mile out of Wallace, there is an old, dirt road that crosses the Union Pacific Railroad tracks called the Hobo Jungle. This same road also went into the back entrance to Baraby the Wrecking Yard.

While driving through Hobo Jungle and the wrecking yard, we had our spotlight going in every direction, and we spotted Lewis's car parked between two wrecked cars. We couldn't see anyone around the car, but both doors were wide open. As we walked up to the car, I could smell a strong odor. On the driver side floorboard, I spotted a .32 special, and my partner noticed that the blond woman's clothing and items were still in the back seat. They had left in a hurry to get away before we could catch them. We called for back-up.

Two deputies from the Sheriff's Office and one Osburn police officer arrived, and we told them what we had going and needed to search through all of the old wrecked vehicles, plus around the South Fork of the Coeurd'Alene River. We told the officers that Lewis may be armed, as we had seen one .32 pistol in his car.

Lewis and the woman were nowhere to be found. We all returned to our patrol cars and started a wider search, beginning at the Myles Motel across from the wrecking yard, but no luck. After an hour or so, I called off the search for Lewis. I told my partner that Lewis was going to have to get to a phone to call someone to pick him up. He may have made it to the backstairs of the U & I rooms.

I wasn't happy we had lost the pair of suspects, and we drove back to the wrecking yard. There, we woke up Baraby (the owner of the yard), and asked him to put a wrecked car behind Lewis's car in case he came back during the night. Before we left the yard, I walked to the rear driver's side of the Cadillac and broke the tail light out. My partner asked me why I did this, and I said, "I should have done this earlier when I was looking at Lewis's license plate. Then the state police or the FBI would have had a good reason to stop the car." My partner just smiled at me.

The next morning around nine o'clock, I called my partner, Doug. I wrote up a report of what had happened the evening before, and we also spoke to our chief. We then went over to the County Courthouse to see the District judge where we got a search warrant to search Lewis's car.

Our search produced a small sandwich bag with a few ounces of marijuana, the .32 pistol, a switchblade knife, the blond woman's clothing, her two suit cases, and a small cosmetic bag. I can not recall if Lewis's mother came to get her car, but I do know it was in the wrecking yard for a week or two with the doors still wide open. All of the items were still in it.

Six days after Butch Jacobson's confrontation with the pimp in Wallace, he shot and killed the cashier at the Bon Marche while committing a robbery.

As I remember the robbery
By (Tony Bamonte)

On October 16, 1971, at approximately 5:00 p.m., I was working an extra assignment to help cover the heavy traffic immediately following the Washington State University/California football game at Joe Albi Stadium. At the time, I was assigned to the traffic department and rode a Harley-Davidson police motorcycle. While directing traffic at the south end of the Monroe Street Bridge, a call came over my police radio that there was an armed robbery in progress at the Bon Marche, later Macy's, and someone had just been shot. I immediately abandoned my assignment and rode to that location. Due to the fact that I was on a police motorcycle and riding against traffic on a one-way street, I was able to arrive within two minutes from the time of the actual murder. I parked my motorcycle on the sidewalk in front of the west entrance to the building (the main entrance).

Due to the knowledge I had just recently received that someone had been shot, I immediately unholstered my pistol, holding it down next to my right thigh to keep it inconspicuous, in an attempt to keep from frightening store customers.

I entered the building at the main entry door on the southwest side of the building and immediately began running up the descending escalator. As I reached either the second or third floor, I observed a man coming down the stairs at a quick pace. At that point, I did not have a description of the suspect and stopped this man at gunpoint. Thinking I was going to shoot him he stated, "It's not me. Follow me – I'll point him out." Believing him, I followed.

We quickly descended down the same down escalator I had just run up. My suspect, at the time, (Michael Yates) led me to the west entry of the Bon Marche, where there was a secondary entry door, which no longer exists. When we reached that location, as I recall, there were several policemen blocking the doorway and also a number of people in the immediate vicinity, especially behind me. (I later learned that my first suspect was the store's security guard, Mike Yates, who later joined the Spokane Police Department, serving over 27 years. I was grateful for his help in locating the suspect and always felt he did his job well.) Within seconds of our arrival at that doorway, my 'then" suspect (the security guard Yates) pointed to the killer, Jerry Lewis, saying, "That's him."

At that time, I still had my pistol inconspicuously down at my side, which I don't think the suspect saw. Within one or two seconds from the time the suspect was identified, and now realizing that, he drew a pistol from his right trench coat pocket, aiming it directly at me. When he did this, there were several other officers around him guarding the door, who weren't paying much attention to him. Because he had been detained for a minute or so without causing any problems or arousing any suspicion, one of the officers mistook him for the janitor, as he was holding a waste basket while he waited there. Also, that fact that he was dressed in what appeared to be casual clothes for the times, allayed any suspicions. As all this was happening, I was cognizant of the inherent danger, due to the number of people in the store. Consequently, there was nothing else to do

FORECAST TODAY
Mostly Fair
Saturday High Low
Airport 50 29
(Full Report on Page 2)

THE SPOKESMAN-REVIEW

89TH YEAR. NO. 156. SUNDAY MORNING, OCTOBER 17, 1971. PRICE 25 CENTS. SPOKANE, WASH.

Valley Woman Killed During Robbery

Murder Charged in Robbery Death

By JIM SPOERHASE

A 35-year-old Spokane man, alleged to have killed the Bon Marche Department Store credit manager Saturday afternoon during an armed robbery, today was charged with first-degree murder.

Pros. Atty. Donald C. Brockett said the murder charge was filed against Jerry Lewis, who police said lives at E3320 Congress.

Detective Capt. Calvin D. Smith said that Mrs. Walter J.

Jerry Lewis

(Peggy Jean) Palmer, 45, E8316 Valley Way, was shot and died almost instantly when an armed man held up the fifth floor business office at the department store.

Lewis, shot in the head by Motorcycle Officer Anthony G. Bamonte as he was fleeing from the store, was reported in critical condition, but showing improvement, today at Deaconess Hospital where he is under police guard.

Held in the County-City Jail under $10,000 bond each as material witnesses were Joe Gibson, 50, who gave an address of E605 Bismark, and James E. Boyd, 27, who said he lived at N2418 Hamilton.

Incident Described

Describing what happened at the store, Smith said:

"Shortly before 5 p.m. on Saturday three men were observed acting in what Bon Marche employes thought was a suspicious manner on the fifth floor, near the credit office.

"An employe notified a store security officer, who was in the act of telephoning police when the incident happened.

"Our investigation indicates that one man confronted Mrs. Palmer with a gun and told her to open the door to the credit office. She said she couldn't open it, that it had to be opened from the inside. Witnesses tell us that the gunman and Mrs. Palmer wrestled a bit—then she was shot."

Capt. Smith said that the robber then jumped over a counter and brandishing the gun—a .22 caliber revolver—ordered three cashiers to help him put money into a pink wastebasket which he picked up in the office.

The employes ordered to hand over the money were identified by homicide Detective Homer C. Hall as Hilda T. Roberts, 56, E1224 Central; Maxine L. Stephens, 52, S6722 Plymouth Road, and Sandra L. Brncick, 22, W1805 Ninth.

"After the robber stuffed the money into the waste basket he went out the office door and down the escalator," Smith said.

By this time, he said, several police officers who had been in the downtown area on traffic duty because of the Washington State-California football game descended on the Bon Marche in response to a call of an armed robbery in progress.

Smith said the armed man got to the first floor of the store and was nearly outside the entrance at Main and Wall when he was confronted by Officer Bamonte.

Gun Is Spotted

Bamonte, who had his gun out and in his hand at his side when he entered the store, said he saw the man grab into the wastebasket and come up with a gun.

Bamonte reported the man had the gun at eye level, aimed at Bamonte when the officer fired. The bullet struck the man in the head and he fell. A pistol was found underneath the wounded man, Smith said.

When the suspect was shot by Bamonte he dropped the money-filled wastebasket which spilled out onto the store floor.

Homicide Detective C. R. Johnson said the cash in the wastebasket amounted to $8,323.

Smith said that soon after Lewis was shot by Officer Bamonte the two material witnesses were taken into custody by other officers in other parts of the department store.

Detective Johnson said an autopsy performed yesterday on Mrs. Palmer's body disclosed that she was shot in the right arm and that the bullet went into her body, killing her almost instantly.

Lewis underwent surgery Saturday night, police said.

Lewis had been scheduled to go to trial next month on a second-deg...

Mrs. Walter J. Palmer

party was allegedly in progress.

No one was hurt in the incident in which several bullets were fired into the house, the Superior Court was told at the time Glover was sentenced.

Philip W. Alexander, managing director of the Bon Marche, said Mrs. Palmer was "a very popular employe and a good friend of everyone in the Bon Marche organization."

"Mrs. Palmer had an outgoing, friendly personality and all personnel of the store feel a tremendous loss and the circumstances of her death have clouded the spirits of her friends," said Alexander.

In addition to her husband, Mrs. Palmer is survived by five children, Judith 24; Mrs. Vickie Louise Taylor, 22; Joseph, 19; Jerry, 16, and Susan, 9; her parents, Mr. and Mrs. Arthur J. Franklin, N1402 Mamer; a sister, Catherine Franklin, Spokane Valley, and brother, Floyd A. Franklin, Sunnyside, Calif.

A Bon Marche employe since 1958, Mrs. Palmer was promoted to credit manager little more than a month ago.

Man Apprehended After Robbery

Police were holding two men as material witnesses Saturday night.

Assailant Shot by Patrolman

(Also see picture, page 6.)

By TOM BURNETT
Spokesman-Review Staff Writer

A cashier's assistant was shot and killed moments before her assailant was shot and critically wounded by a police officer during an $8,000 armed holdup Saturday evening at the fifth floor credit offices of the Bo... Marche, Main and Wall.

The victim was identified as Mrs. Peggy J. Palmer, 47, E8616 Valley Way, mother of five children.

The two shootings occurred at about 5 p.m., a prime shopping time for the busy downtown department store, and hundreds of customers were inside the store.

Witnesses said three men e... tered the store, walked to th... elevator and rode to the fift... floor. The elevator is locate... near the rear of the large d... partment store, leading police ... believe that the men were insid... for some time before using th... elevator.

Two Go to Office Area

At the fifth floor, one man remained at the elevator while the other two men walked into the "cash office" area. The cash offices — five teller windows and a credit office — are located about 100 yards from the elevator, through the store's toy section.

Police said one of the men in the cash office area sat on a bench in the area and was not involved in the actual holdup.

The second of the two men — the man who eventually was shot — proceeded into the cashier credit office. The office is staffed by three persons. However, only one — the victim — was on duty at the time of the robbery, police said.

The man reportedly grabbed Mrs. Palmer and forced her into a small area behind her desk. The shooting apparently occurred in that area, police said.

One shot was fired at Mrs. Palmer, who fell to the floor mortally wounded.

After the shooting, the gunman walked from the credit area to the cash offices and approached one of the five windows. The man leaped over the counter, grabbed a small plastic waste basket and stuffed it half full with paper money, police said.

Others Unaware of Incident

Store employes in the immediate area were totally unaware of the incident until the fatal shot was fired.

After the gunman took the $8,000, he jumped over the counter and out into the customer area. The man who had been seated on the bench and the gunman then walked swiftly away.

They were joined by the third man, and the trio then dodged store customers as they ran down the four flights of escalator stairs to the first floor.

...e Bon Marche store detec... who arrived at the scene of ... shooting moments after it ...ened, followed the three ... to the first floor.

... the time of the incident, police officials were assigning patrolmen who had handled the heavy traffic after the Washington State-California football game to downtown areas for the remainder of the evening.

At least 50 policemen were in the general downtown area when the shooting of Mrs. Palmer occurred, and about 40 officers went immediately to the store after the initial alarm went out over the police radio network.

As the trio neared the first floor, Police Sgt. Gene McGougan, who arrived within one minute after hearing the alarm, was riding the escalator to the second floor.

...en Told to Stay in Store

... the three men neared the ... Street entrance, Patrolman ...ld B. Graves was coming ... the store. He ordered the ... to "remain in the store" ... not try to leave.

...aves said the men turned ...d and walked a few feet to ...nt where they were about ...t from him.

...rolman Anthony Bamonte, ... had entered the store ...gh the main entrance, ar... at the scene.

... the men turned, they con...ed Bamonte.

... man carrying the plastic wastebasket reached into the receptacle, produced a revolver and pointed it at Bamonte, the officer said. Bamonte pulled out his service revolver and fired once at the man, with the bullet striking the man in the head.

The man was reported in critical condition at Deaconess Hospital.

As the man fell, the wastebasket landed on the floor and the money spilled onto the stairs. Sunglasses the man had been wearing later were found shattered a short distance away.

...air Held by Authorities

... two other men were ar...d and held as material wit...s Saturday night.

...en the shooting occurred, ...wo patrolmen and the three ... were by themselves at the ... entrance except for a ... girl who was making a ...one call nearby.

... customers or employes ... in any immediate danger, ... said.

...arge crowd of sightseers ...red at both store entrances ... the incidents. Police set up ...icade to seal off the store.

...ce then searched the en... normal Saturday night closing time. All employes were escorted out of the building through a rear door, except for those who work in the areas where the incidents had occurred. They remained to answer detectives' questions.

Deputy Chief of Police Thomas J. O'Brien took charge of the investigation, assisted by Capt. Calvin Smith. County Pros. Atty. Donald C. Brockett and Asst. Pros. Atty. James Krum were called to the scene as was County Coroner Dr. William E. Jones.

The above headline and article are from the October 17, 1971, *Spokesman-Review*. The motorcycle officer shown with one of the handcuffed suspects, was Gary Lacewell, who had just identified and arrested the second suspect.

There were three people involved in this robbery. Two of them were arrested at the scene. The third shot and killed the cashier (Peggy Palmer) approximately three minutes before he attempted to shoot me.

The partial article on the left appeared in the October 18, 1971 *Spokesman-Review*. It contains a photo of the murderer, Jerry Lewis, and Peggy Palmer, the victim.

but shoot him before he shot me. More important, if I had hesitated and given him the chance to shoot first, he would either have shot me or, because that section of the store was filled with people, there was a high probability of his killing or wounding one or more store customers. The identification of the suspect by the security person and the shooting process lasted about two seconds.

As there was no time to process my actions, and with my pistol still down at my side, I raised it slightly and fired one shot from my hip level. Lewis was wearing sunglasses and my bullet struck the glasses directly on the bridge, breaking them in two. Because he was aiming his pistol at me, his head was turned sideways. After entering the middle of the perpetrator's forehead, the bullet exited the left side of his head. Instantly, he fell forward and onto his pistol. Having his head turned to the side saved his life. At the same time the wastebasket he was holding fell to the floor, spilling out much of the $8,323 it contained. Immediatley, after this happened one of the officers standing right next to him, still thinking he was the janitor, made the statement, That F... ing Bamonte just shot the janitor.

Butch Jacobson's suspect, Jerry Lewis, lived but was in a wheelchair for the rest of his life. According to Chief Wayne Hendren and Spokane County Prosecutor Donald Brockett, Lewis was the first person in the history of Washington State to ever plead guilty to a first degree murder charge. However, this was a technicality, as at the time he pled guilty he was in Eastern State Hospital, which is a psychiatric hospital at Medical Lake (he lost about a third of the frontal lobe of his brain). Once he had recovered sufficiently, it was far cheaper to keep him at the Washington State Penitentiary which is a Washington State Department of Corrections men's prison located in Walla Walla, Washington.

As, it was during a time of a lot of racial tension, and the fact the murderer was black, the shooting created a bit of controversy. As a result, the Police

In 2009, Judy (Palmer) Bendewald, authored a book about Elvis (left photo), which included her many encounters with him. From 1966 to 1972, Judy was the president of the "Kissin Cousins" Elvis Presley fan club in Spokane. Her book is titled, *My Treasured Memories of Elvis*, Published by Memphis Explorations, in Memphis Tennessee. Judy's book is a great read and a short lesson in Spokane history. Judy was inspired to write her book following a chance encounter with Carla Savalli, at the time, assistant managing editor for the *Spokesman Review* – also an Elvis fan. Right photo, left to right: Sandy Schmidt, Elvis Presley, and Judy Palmer. *(Courtesy Judy Palmer)*

ated a bit of controversy. As a result, Spokane Police Chief, Wayne Hendren, immediately contacted the media and told them the circumstances. The knowledge that the murderer had gunned down the mother of five children, then attempted to shoot a policeman, was all the Black community needed to reach the conclusion that the shooting was not racially motivated. Also, the fact that I only fired one shot, as needed, clearly indicated I only did what was necessary.

Following that incident, I went back to work feeling fortunate there were no other victims. Above all, it was terrible for the victims and families. It was a tragic incident that left grim memories for many people.

Almost 40 years after Peggy Palmer, the cashier at the Bon Marche, was shot and killed by Jerry Lewis, I had the occasion to meet her oldest daughter, Judy (Palmer) Bendewald. At the time her mother was shot, Judy was the president of the International Elvis Presley Fan Club.

Shortly after Judy's mother was killed, Elvis and his wife sent her a sympathy card, which is among her most cherished possessions.

When I met her, Judy had many questions that had remained unanswered since her mother's tragic death. One of them was whether her mother had done anything to provoke this man into shooting her. My answer was, "No, she did everything she was asked to do, including giving him $8,000." There was absolutely no reason for him to have shot her.

A Dangerous Town – "Overkill by the FBI"

The Federal Bureau of Investigation Raids the Little Town of Wallace The *New York Times* Assigns Timothy Egan to Write the Story

Gambling Raid Angers Mining Town

150 agents seize 200 video poker games in bars.

By TIMOTHY EGAN
Special to The New York Times

WALLACE, Idaho — Well before Idaho was a state, there were gambling houses, brothels and boisterous saloons in this storied mining country in the Coeur d'Alene Mountains.

So it wasn't exactly a surprise when, on the morning of June 23, Federal agents from all over the West raided virtually every bar here in Shoshone County and found more than 200 video poker machines. Gambling, local officials say, is one of the few industries that have yet to die in the area.

What is so perplexing to residents of the panhandle of north Idaho, and to outsiders as well, is why the Federal Bureau of Investigation used such a show of force. The raiding party on that Sunday comprised 150 agents, and now Federal lawyers are moving in court to take control of the places that were raided, a move that would make the Government owner of every bar in Wallace.

To a community that considers itself on its knees, racked by a series of economic and environmental calamities, the raid has provoked protests and stirred old animosities.

Machines Not Target, U.S. Says

In an age when banking scandals have cost the nation's taxpayers billions of dollars, many residents here say the Government has spent far too much time and money on video poker machines in a crippled mining town.

Federal officials say the focus of the raid was not so much the machines as the local authorities who had allowed them to flourish.

The raid was the biggest single Federal law-enforcement raid ever in the Rocky Mountain region, said James T. Screen, a spokesman for the F.B.I. in Salt Lake City. Not since the late 19th century, when Federal troops were sent here to battle union organizers, have so many Government agents moved so heavily against one community in the region.

Residents of this valley, which once produced more silver than any other place on earth but is now poverty-stricken and polluted, say the raid can be explained more by the area's reputation than by any real criminal enterprise.

Panhandle Goes Its Own Way

"Here you've got a community that's always lived a little bit outside the law," said Chris Stuecker, a local contractor who specializes in renovating

An article about the infamous FBI raid on Wallace, which appeared in the *New York Times*. It was written by award winning journalist Tim Egan, and published on August 20, 1991. *(Public domain)*

Wallace, Idaho, Makes the *New York Times*

I have known Tim Egan since 1989, when he interviewed me a number of times prior to writing the book *Breaking Blue*, which was about a case I investigated. I have found Egan to be somewhat of a champion for the underdog. He recognizes and often writes about things that are unjust and often corrupt involving public officials. Tim Egan recognized the manner in which the FBI conducted their raid on Wallace. If the FBI felt what was going on in Wallace was a problem, as a former law enforcement officer, I know it could have been handled in a much fairer and evenhanded way. The way this was handled was a massive overkill. What they intended to accomplish could have easily been done by making one or two arrests to get the word out that this type of activity was no longer going to be tolerated.

An article on this raid appeared in the *New York Times* on August 20, 1991. It was written by Timothy Egan, a writer for that newspaper.

Gambling Raid Angers Mining Town

By TIMOTHY EGAN,
Published: August 20, 1991

WALLACE, Idaho— Well before Idaho was a state, there were gambling houses, brothels and boisterous saloons in this storied mining country in the Coeur d'Alene Mountains.

So it wasn't exactly a surprise when, on the morning of June 23, Federal agents from all over the West raided virtually every bar here in Shoshone County and found more than 200 video poker machines. Gambling, local officials say, is one of the few industries that have yet to die in the area.

What is so perplexing to residents of the panhandle of north Idaho, and to outsiders as well, is why the Federal Bureau of Investigation used

Timothy Egan, *New York Times* writer, circa 2005. *(Public public domain)*

such a show of force. The raiding party on that Sunday comprised 150 agents, and now Federal lawyers are moving in court to take control of the places that were raided, a move that would make the Government owner of every bar in Wallace. To a community that considers itself on its knees, racked by a series of economic and environmental calamities, the raid has provoked protests and stirred old animosities.

Machines Not Target, U.S. Says

In an age when banking scandals have cost the nation's taxpayers billions of dollars, many residents here say the Government has spent far too much time and money on video poker machines in a crippled mining town.

Federal officials say the focus of the raid was not so much the machines as the local authorities who had allowed them to flourish.

The raid was the biggest single Federal law-enforcement raid ever in the Rocky Mountain region, said James T. Screen, a spokesman for the F.B.I. in Salt Lake City. Not since the late 19th century, when Federal troops were sent here to battle union organizers, have so many Government agents moved so heavily against one community in the region.

Residents of this valley, which once produced more silver than any other place on earth but is now poverty-stricken and polluted, say the raid can be explained more by the area's reputation than by any real criminal enterprise.

Panhandle Goes Its Own Way

"Here you've got a community that's always lived a little bit outside the law," said Chris Stuecker, a local contractor who specializes in renovating historic houses. "And then you get an outside agency, like the F.B.I., that doesn't understand that we're just average folks, and they hear about all these gambling devices and it gets them going." Mr. Stuecker helped to organize a rally last month to protest the raid, and about 200 people attended.

This town of about 1,000 people is a national historic site, an example of the boom-and-bust mining towns that used to dot mountain valleys throughout the West. Even the newer buildings here look archaic. But, Federal officials say, it is not just architecture that is frozen in time in the Silver Valley: so are attitudes.

Many residents and some public officials of the valley admit that their county is different from most. Idaho is better known for potatoes and Mormons than gambling and prostitution. But the northern part of the state, in the mountainous panhandle, has always gone its own way.

When the silver mines were running rich, the area was full of rough-edged men who spent their earnings on the basic vices of typical mining and timber camps – gambling and prostitution. Once, more than a dozen brothels operated in Wallace, local residents say.

But the Silver Valley has changed dramatically in the last decade. As the mines have closed and toxic levels of lead have been found in local streams and soil, Shoshone County has lost more than 27 percent of its residents, down to about 14,000 now. Unemployment has been running at more than 20 percent, and last year the last known brothel went out of business. "It died a natural death," said Dick Caron, who owns the Wallace Corner, a general store.

'Corruption of Public Officials'

Federal agents say local authorities have long tolerated the traditional forms of vice here, and the raid was intended to end that. "Corruption of public officials was the focus," said Mr. Screen of the F.B.I.

The case is being handled by the Justice Department's Division of Public Integrity, a unit in Washington, D.C., used primarily to prosecute corrupt elected officials. Department officials would not comment on the investigation.

Grand jury proceedings are under way in Boise, but no one has yet been indicted. Agents confiscated $500,000.

The 150 F.B.I. agents raided 58 bars, breaking down doors and windows in some cases. So many agents were needed, Mr. Screen said, because there were so many bars.

The agents confiscated more than $500,000 in cash from various bars and seized all the video poker machines. The machines are legal in Idaho but are considered unlawful by the state if they are used for actual gambling – that is, if the bar pays players for points scored on the machines. 'we're talking quarter games."

Rather than expressing outrage at crimes that the authorities say were going on in their midst, Wallace residents have directed their anger at the F.B.I.

"What we have in these saloons is no more harmful than bowling on Sundays," Mr. Caron

said. "It's petty, insignificant, victimless entertainment. We're talking quarter games."

Jack Rose, the Shoshone County Prosecutor, said, "I think the F.B.I. would be better served trying to solve the many killings back in their headquarters city, the murder capital of the nation."

Mr. Rose said that he had done nothing wrong and that he was not aware of any gambling on the video poker machines.

F.B.I. agents, who spoke on the condition of anonymity, said Mr. Rose was one of the public officials under investigation. Justice Department officials refused to comment.

T-shirts criticizing the F.B.I. quickly sold out at Dickison's Trading Post here. A local newspaper, The Silver Valley Voice, published a satirical front-page picture of a clownish-looking man, posing in fake nose and glasses, wearing an F.B.I. hat and shorts while standing in front of a bar. Several hundred people have signed a petition calling for a Congressional investigation of the raid; it accuses the agency of selective enforcement of the law.

'Colorful Way of Life'

Video poker machines can be found throughout the state, said Keith Mathews, assistant director of the Idaho Department of Law Enforcement. "I don't think that Shoshone County is any different than any other part of Idaho," he said.

The area's Representative in Congress, Larry LaRocco, a freshman Democrat from Boise, said the raid left him with many questions.

"I wonder if they are using an F-117 when a small rifle would do the job," Mr. LaRocco said, adding that the Silver Valley has long had a "live and let live" attitude and "colorful way of life" that local officials have tolerated.

Mr. Caron, the merchant, said, "We might be hicks in the sticks, but we know what overkill is."

It was unfortunate the way the FBI pulled off that bust, but they learned from it. They learned that high-handed tactics don't work. They made a major blunder in their entire operation and in the long run they were not successful. If they wanted to accomplish their goal, it would have been far easier to make an arrest or two with the worst offenders, and soon they would have accomplished what they wanted.

The brothels did not close down because of the FBI raid. They closed down because of the poor economy due and the AIDS scare.

Wallace Was Always a Town of Corruption

– In the Eyes of the Law –

The 1929 Raids Nab Corrupt Wallace Officials

And Clean-up the Town – For a Short Time

"Prohibition Raids Anger Mining Towns"

The 1929 Prohibition Raids in Shoshone County,

Similar to the 1991 Illegal Video Poker Games Raided in Wallace.

Law Enforcement Used to be Much Harsher on Lawbreakers.

Fortunately, during the 1991 FBI raids the Feds were not allowed to shoot people who owned video poker games.

During prohibition the production and sale of alcohol was a felony offense, over which distillers and liquor traffickers could be shot and killed. Some were even shot in the back as they tried to flee the law.

Today, things are much different. Liquor manufacturing and regulation is a profitable governmental controlled function. Government authorities have learned that by controlling it and having a hand in its sales, it is highly profitable. They now approach this former problem with remarkable enthusiasm, and as a result the states now control the sales of alcohol themselves. They are now doing the same thing that, at one time, they allowed law enforcement officers to shoot people who committed the crime of making and selling alcohol.

Dry Agents Ordered to Stop Shooting Victims

Too Many Fatalities As Result of Itching Trigger Fingers of Still Raiders

WASHINGTON, June 4.—Gun battles between federal dry agents and bootleggers, moonshiners and rum-runners resulted in 140 casualties since prohibition became effective, the United Press learned today from official figures.

Because of the growing fatality list Dry Czar Andrews has ordered his men to use weapons only in emergency and shoot only in self-defense.

Forty-nine prohibition officers have been killed in the line of duty, while 92 persons resisting or evading arrest have been killed. Most of the slain officers lost their lives while raiding stills, although several were murdered to prevent their giving evidence at trials.

Idaho was one of several states, along with Washington and Oregon, to enact statewide prohibition on January 1, 1916, four years ahead of when the National Prohibition (Volstead) Act went into effect. At the time, much of the state was already dry by authority of local county or city ordinances, but most residents of Shoshone County were not supportive of the new act and made little effort to hide their contempt and defiance of it.

Miners and lumberjacks had long-standing reputations for their liberal consumption of alcohol, and those in the Coeur d'Alenes did their parts to uphold that tradition.

The demand for alcohol from a large number of young, single, high-spirited men working in the woods or the district's mines ensured readily available sources. Though Idaho's statewide law is claimed to have been among the strictest in the country, it did little to dry up the illegal liquor supply in Shoshone County.

Retrospective studies regarding the effectiveness of Prohibition have concluded it was a miserable failure on many levels, creating far more problems than it was ever designed to solve. Historians have concluded that Prohibition created a nation of drunkards and criminals. There was also rampant hypocrisy, as many in enforcement positions were known to be heavy imbibers.

Though Shoshone County was not much different than countless other regions in its violation of the Prohibition Act, it made headlines nationwide after a series of raids and numerous arrests in Wallace, Mullan, and Kellogg in 1929.

Those small cities captured much attention due to the large number of county and city public officials who were arrested and accused of widespread conspiracies to condone, protect, and illegally benefit from illicit liquor trade. During their wholehearted cleanup, federal undercover agents arrested nearly two hundred residents, including Shoshone County

"Prohibitionists in a Dry State." Dr. Leonard Hanson, of Wallace, Idaho, drinking whiskey at a friend's home on the Two Mile Road in 1920. Hanson's nephew, Irving Anderson Jr., is at the far right. Many people scoffed at the laws prohibiting the production and sales of liquor, giving rise to the term "scofflaw," which was coined during Prohibition. *(Robert L. Anderson collection)*

Sheriff Rene E. Weniger; sheriff deputies, Charles Bloom and Albert Chapman; Wallace Mayor, Herman J. Rossi; Shoshone County Assessor, William H. Herrick (also a former Wallace mayor); and high-ranking police officers, including the chiefs, of Wallace and Mullan. Most of the arrested were found guilty and handed federal prison sentences, along with hefty fines.

However, within the next two years, a large percentage of the cases were overturned during the appeals process. Allegations of organized conspiracy could not be adequately proven, and the city and county officials had not profited personally by their acts.

Lawyers successfully argued they had operated "for the good" of their county or communities. Their acts involved the collection of license fees (essentially amounting to fines) from saloons (thinly disguised as soft-drink parlors) and other illicit establishments, such as gambling halls and houses of prostitution. Some were also known to have accepted bribe money to look the other way (an offense not so readily viewed as "for the good of the community"). The officials involved justified their actions by the effectiveness of the fines system, which helped to cover municipal projects and defray some of the additional costs imposed by the Volstead Act. Prohibition had stamped out the considerable revenue the legitimate liquor licenses had previously produced. For elected officials to interfere with this new source of revenue, or to take a hard line in enforcing the bans on liquor trade, would likely have been political suicide.

Many people scoffed at the laws prohibiting the production and sales of liquor, giving rise to the term "scofflaw," which was coined during Prohibition.

The system of fines gained the support of the local citizenry, who flocked to the trials in a show of solidarity – sometimes lending a colorful circus-like atmosphere.

A statement of support toward Weniger and Herrick was also shown during the primary election in August 1930. Their wives ran for the positions they held prior to conviction. Ethel Weniger won the Democratic nomination for sheriff but was defeated in the general election by Walter Hendrickson. William Herrick's wife, Margaret A. Herrick, received the Republican nomination and became the next county assessor. While his wife was campaigning for office, Herrick was serving time at the federal prison on McNeil Island in Washington State.

There was such strong community support that an article in the *Spokesman-Review,* "wagging a finger at Wallace for being a "modern Sodom," created intense anger and lingering resentment toward Spokane.

After pointing out that Spokane had been built on wealth from the Coeur d'Alenes, about a hundred Shoshone County businessmen vowed to stop their subscriptions to Spokane's newspapers and boycott businesses that supported the papers. No doubt it was not lost on them that Spokane was also steeped in its own Prohibition-related corruption and blatant violations of the liquor laws.

The courts eventually ruled it was not illegal to collect fines against or tax "that which is forbidden." When, during appeal, the courts overturned cases of some of the higher-ranking officials, Hoyt E. Ray, the U.S. district attorney who led the initial campaign in what has been called "North Idaho's Whiskey Rebellion," threw out some of the other cases. Although both Sheriff Weniger and Mayor Rossi resigned their respective positions, and were sentenced to serve time at McNeil Island Prison, in 1931 their cases were overturned in appeal and neither served time. Charles Bloom's case was also overturned. In 1934, the year after Prohibition was repealed, many received full pardons from President Franklin Roosevelt.

The Spokane Police Department's "Dry Squad," circa 1925. Spokane County Motorcycle officer Bob Bean's grandfather was a member of the Spokane police department for 25 years. During the time of prohibition in Spokane, as a detective, he was assigned to the "Dry Squad." Prior to Bob's mother's death (June Bean), Bob told me (Bamonte) I should contact her as she had some family photos involving the police department. She actually didn't have many police related photos, but provided the above photo with an interesting story:

The Dry Squad was specifically designed to enforce the prohibition of alcohol when Prohibition went into effect nationally between 1920 and 1933. During that time, June's father, Walter Hubert, pictured third from the left, joined the force in 1922 and retired in 1947. As a member of the Dry Squad, Hubert was often required to work with a partner. He later confided to his daughter he hadn't enjoyed working with his partner because "he was a heavy drinker and to cover the alcoholic aroma, he'd eat garlic before coming to work and smoke cigars in the car. Somehow the combination produced a lot of gas, which didn't make for pleasant working conditions. *(Bamonte photo from June Bean)*

The Book *Breaking Blue* by Timothy Egan And its connection to Wallace

Spokane Police Chief Terrance Mangan Gets Caught in a Lie which Appears in Egan's Book – *Breaking Blue*

***Breaking Blue* is a book about a Spokane Police Detective, the key person in a burglary ring, who shot and killed a Newport, Washington marshal. He commited the murder during a burglary he was committing. The murder was covered up by three different law enforcement agencies.**

In his book, Tim Egan identifies Chief Mangan as lying about this case. Also, in the Epilogue of Breaking Blue, on page 253, Egan makes the statement:

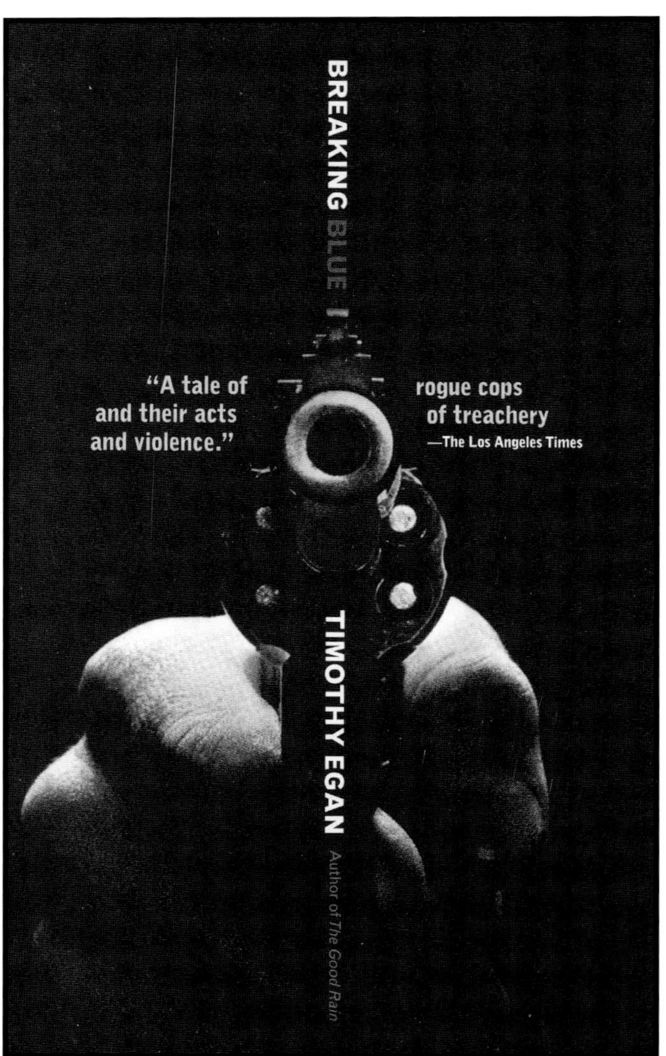

The book *Breaking Blue* by New York Times writer Timothy Egan. *(Bamonte photo)*

A photo of Bill Parsons (left) and Dan Mangan given to Sheriff Tony Bamonte by Bill Parsons to prove they were working together. Parson's and Mangan were the police officers that threw the murder weapon into the Spokane River. I drove each of them separately to the bridge and they both marked the same location where in the river I would find it. circa 1935. *(Courtesy Bill Parsons, donated to Bamontes)*

> ***As the election neared, the Spokane police chief called three press conferences at which he specifically criticized Bamonte. Some reporters were baffled. What was the chief of the biggest police department in the region trying to do? They had never seen one cop bring the press together to ridicule another lawman. On primary day, September 18, Sheriff Bamonte lost his job by 34 votes.***

Prior to becoming the Police Chief of Spokane, Mangan had been a Catholic Priest. His priesthood only lasted for a few years when he made the decision to forsake and violate the vows he had made when he became a priest. He now decided he wanted to become a police officer.

Over time, as a law enforcement officer, he had on numerous occasions proven to be dishonest and vindictive. The book, *Breaking Blue*, by Timothy Egan brought out the worst in Mangan's char-

Murder victim George Conniff, circa 1930.
(Courtesy Conniff family)

acter. What Chief Mangan did as a payback for catching him in a lie, was vicious and unjust. It would also falsely imprison a man for life.

To the date of this book, Egan has written eight books. One of his books won the National Book Award, *The Worst Hard Time: The Untold Story of Those Who Survived the Great American Dust Bowl*. The National Book Award was established in 1950, and is an American literary prize administered by the National Book Foundation, a nonprofit organization. A pantheon of such writers as William Faulkner, Marianne Moore, Ralph Ellison, John Cheever, Barnard Malamud, Philip Roth, Robert Lowell, Walker Percy, John Updike, Katherine Anne Porter, Norman Mailer, Lillian Hellman, Elizabeth Bishop, Saul Bellow, Flannery O'Connor, Adrienne Rich, Thomas Pynchon, Isaac Bashevis Singer, Alice Walker, Charles Johnson, E. Annie Proulx, and Timothy Egan have all won the Award.

Egan's first book, *The Good Rain*, won the Pacific Northwest Booksellers Association Award in 1991. In 2001, the New York Times won a Pulitzer Prize for National Reporting for a series to which Egan contributed, "How Race is Lived in America." Egan currently lives in Seattle, and writes a weekly op-ed for the *New York Times*.

The Story of Breaking Blue And how it came about

In 1989, I had been attending Gonzaga University finishing a masters degree. My theses was on all the sheriffs of Pend Oreille County and the murders each had investigated during their tenure in office. As a result, I came upon the murder of a Newport, Washington marshal by a Spokane policeman.

In 1987, during the time I was the elected sheriff of Pend Oreille County, I entered a masters program at Gonzaga University. My degree was Organizational Leadership. When the time came to write a thesis, I approached my advisor and suggested it be on the leadership styles of each sheriff in Pend Oreille county, where I was then the eleventh sheriff. The concept was met with approval. Consequently, I began the study of each sheriff that held office and the major crimes they confronted during their tenure, and how they were solved.

In 1989, my thesis was almost complete and I was ready to defend it, which was the final phase toward getting a degree. About the same time, the Gonzaga *Sigma*, a small Gonzaga publication, which is no longer active, ran a story about what they considered some of the more interesting theses. Mine was among them. As a result, a reporter from the *Spokesman-Review* read their story and also did their own story. Following that, Timothy Egan from the *New York Times* saw the *Spokesman's* story, which was on the *Associated Press,* and wrote a piece about my thesis, which appeared on the front page of the *New York Times*.

When the first story in the *Spokesman-Review* came out, the date of the murder was not correct. That same afternoon, the son of the murdered marshal, George Conniff Jr., contacted Bill Morlin, an investigative reporter from the *Spokesman-Review*. Conniff Jr. wanted to know more about the "shooting of the marshal" story. He stated it sounded like his father, who was shot in 1935, but the dates on the *Spokesman* appeared to be wrong. Morlin started looking up information and soon learned the dates of events in the first *Spokesman* story were wrong. He confirmed to Conniff Jr. that the story Camden wrote was about his dad, George Conniff. He then gave him my number and asked him to call.

From that point on Morlin became involved. During his years of working for the *Spokesman-Review*, he had developed many contacts. As a result and concerning this case, he consistently was being contacted by people who had been involved in or who knew about the Conniff murder and had information to give, which he always passed to me.

As a result of his help, I was able to develop what would turn out to be the nation's oldest active murder case. Had Morlin not taken an interest, and not developed numerous leads for me, this case would not have been solved. *Breaking Blue*, a *New York Times* best seller, was published about this case in 1992.

The Story of *Breaking Blue*
SETTING THE SCENE FOR SOMETHING FAR WORSE

SPOKANITE WILL TELL HER STORY 1935 NEWPORT MURDER FEATURED ON TV SHOW

By Bill Morlin, Staff writer

For about 50 years, no one has really listened to what Pearl Keogh knows about the 1935 murder of the town marshal in Newport, Wash.

A year after the killing, Keogh said she and her sister gave authorities information and evidence, but were branded as "just a couple of women who should mind their own business."

Tonight, the spunky 85-year-old Spokane Valley grandmother will tell her story on prime-time TV to 28 million Americans.

"I'd just like to see justice done, somehow, but I don't know how it's going to be done now, after all these years," Keogh said Monday between sips of coffee at a Spokane Valley restaurant. "I feel so bad to think that something couldn't have been done before now," she said, gazing out the window.

Keogh contacted Pend Oreille Sheriff Tony Bamonte last year after reading news accounts on the unsolved slaying of Newport Marshal George Conniff. It wasn't long before the TV cameras of "Unsolved Mysteries" showed up.

Keogh said she's convinced former Spokane police detective Clyde Ralstin was involved in the death of Conniff. He was gunned down after startling burglars stealing dairy products, then scarce, from the Newport Creamery on Sept. 14, 1935.

Ralstin, now 90 and living in Montana, has denied involvement in the killing. He has hired a lawyer, who says Ralstin is too old and ill to talk with investigators.
Keogh said her belief is based on a confession

Pearl Keogh, circa 1935. *(Courtesy Pearl Keogh)*

The Spokane River as it looks without the water flowing over it. From left to right in this picture are members of the Spokane Treasure Club, several members of the media, and myself. The Post Street Bridge is to the right of this photo and in the direction the group is heading.

This case, which was on *Unsolved Mysteries*, was the beginning of hard times for me as an elected sheriff – both from the *Spokesman-Review* and Chief Terrance Mangan of the Spokane Police Department. Photographer, Dan Pelle, from the *Spokesman-Review*, was assigned to take pictures of this. Probably, because the river had just been drained and we could be seen slipping around while walking on the rocks. He was watching us while sitting on the Post Street Bridge railing, where he remained. Once the pistol was located, I left it in place, for about 30 minutes, so pictures could be take of it. I then cleaned it off and the entire search group headed back up to our staging area, which had been the Black Angus parking lot (now Anthony's Restaurant). When we got to the staging area, Mr. Pelle approached me and told me he needed me to take the pistol back down where I had found it, and put it in the spot I found it so he could take a photo of it. I advised him that I had a photographer that had already taken pictures of it (see following page) and he could use one of those. He stated it had to be a picture that he had taken. I then told him that if I staged the photo it would be obvious that it was a staged photo and I wasn't going to do that. Angry words soon followed from both parties involved, and he left. Later I received a call from someone at the *Spokesman-Review*, who asked if I would let someone else take a photo of the pistol with me holding it. I agreed to this, which they did.

This was the case that Spokane Police Chief Terrance Mangan lied about, was caught lying by *Spokesman-Review* reporter Bill Morlin, and exposed in Egan's book. This was the beginnings of my bad experiences with the Spokane police chief. *(Photo by Lynda Donoian)*

The murder weapon as it looked after lying in the Spokane River for over 54 years. The handle grips were completely rotted off. Also, the hammer, trigger, and trigger guard were rusted off. When I sent it to the Washington State Patrol crime lab, they stated it was the same caliber Marshal Conniff was shot with and its condition was consistent with being in the river for over 50 years. A Washington Water Power official also stated, "The location where it was found was the only area in that vicinity that had never been disturbed." Last, but not least, it was within 30 feet downstream of where Dan Mangan stated he dropped it. That location was confirmed by Bill Parsons, a former police chief for Spokane, who was with Mangan when he dropped it from the Post Street Bridge railing.

This case was being closely followed by the news media. About a week before the river was stopped by Washington Water Power (now Avista) for surveying, I contacted the Spokane Police Department and the news media to let them know the time and date I would be searching the river. I also invited the SPD, as it was a crime one of their men was a suspect in. However, since Chief Terry Mangan (no relation to Dan Mangan) had publicly made the false statement that Clyde Ralstin had never been on the Spokane Police Department, he failed to send anyone to the scene. I also invited the Spokane Treasure Club to the site, as I had no idea if the river bed would be covered with silt or debris.

When that date arrived and the river was stopped, within 15 minutes I found the pistol exactly as shown in the above photo. After leaving it at the site of discovery for that amount of time for the news media to photograph it. When I arrived at my office in Newport, I placed it in the evidence locker. Once the suspect had passed away, I donated it to the Northwest Museum of Arts and Culture in Spokane. *(Photo by Lynda Donoian)* *she was given in 1940 by the late Virgil Burch, and butter wrappers from the Newport Creamery she found after the shooting in Burch's former home and the "Mother's Kitchen" restaurant in downtown Spokane. Burch, who was both a part-time plumber and restaurant co-owner, had done plumbing work at the Newport Creamery in the weeks before the shooting and may have been recognized by the town marshal, Keogh said.*

She recalled hearing police officers discuss "and even joke about" the Conniff killing at the restaurant at W. 24 Riverside, where they'd stop off for free coffee and pie. She often frequented the restaurant. Her sister, the late Ruth Coffman, worked there as a cook for Burch.

"They talked about it all the time" in the restaurant, Keogh recalled. "Everybody would come in and say, 'Well, who did you kill today?' "They were laughing and made a real laughingstock out of it," she said, "and that's what made my sister so disgusted that she finally quit."

Police complicity in the killing and subsequent cover-up were common knowledge, Keogh said. "There was no doubt in our minds," Keogh recalled. "Every place you went in Spokane, somebody knew about it and who was involved."

Mother's Kitchen bought its dairy products from Hazelwood Dairy in Spokane, but about the time of the shooting, she and her sister found Newport Creamery butter wrappers in the restaurant's garbage. Keogh also found Newport Creamery wrappers in the small North Spokane home once occupied by Burch. In the spring of 1936, Keogh and her sister took the butter wrappers and the information they had overheard in the restaurant to Spokane County sheriff Ralph Buckley. He promised to investigate but never did, she said.

In 1940, after she and her husband had moved to Portland, Burch came to her home for dinner, Keogh recalled. Over dinner, Burch said that he, Ralstin and the late Acie Logan, who had been implicated in other creamery burglaries, were

On the night of September 14, 1935, former Spokane Police Motorcycle Officer, now detective Clyde Ralstin, while committing a burglary of a creamery in Newport, Washington, with two other men, shot and killed Marshal George Conniff. Several years earlier, as a police officer, he had shot and killed a 15-year-old boy from Rogers High School. *(Public domain)*

present when Conniff was shot outside the Newport Creamery, she said. "I was sure of it before," but Burch didn't admit his involvement until the dinner meeting, Keogh said. When she saw Ralstin in the restaurant and asked him about the matter, he always changed the subject, she said.

Newspaper archives and county records reveal Ralstin fatally shot a fleeing Spokane teenager in the back in 1937. A coroner's inquest ruled the shooting was justifiable.

Most police department personnel records have disappeared or can't be located, but a letter obtained by The ***Spokesman-Review*** *and* ***Spokane Chronicle*** *suggests Ralstin left the department while on a suspension three months later. He later became a supervisor at an atomic bomb plant and eventually became a municipal court judge. "It's all hogwash," Ralstin said last August when questioned about his alleged role in the shooting. He has hired an attorney and refused to talk with investigators.*

But still he remains a pivotal figure in what is being called the oldest, active unsolved murder case in the United States – one that is getting national media attention. It will be the lead segment at 8 tonight on the NBC show "Unsolved Mysteries."

George Conniff's children – 54 years after their dad was murdered by former Spokane Police Motorcycle Officer Clyde Ralstin. From left to right, Mary, George Jr., and Olive. The hardest part for them was when they found out Darrell Holmes, the Pend Oreille County sheriff from 1942 to 1953, was part of this cover up. In 1953, Holmes resigned as sheriff and was appointed to the position of a United States Marshal for the District of Washington, serving for eight years. He finished his law enforcement career as chief jailer of Spokane County. This photo was taken the day they learned their dad's murder had been covered up by many law enforcement agencies who knew the facts. Their father's death and the cover-up of his murder had a lasting effect on their lives. *(Bamonte photo)*

The Murder of a Spokane Newspaper Carrier in 1987, and Its Ties to the Wallace Brothels

Have you ever thought about being put in prison for a crime you didn't commit, and unsure of what lay before you? When you are in prison, all personal control will been taken from you. What you wear, what you eat, where you sleep, and where you go, are all dictated by the prison authorities. In prison, the absolute power of government completely controls you and your every move. Or, can you imagine being beaten to death and the top people in law enforcement doing their best to conceal evidence that would prove who your killers really were? That's what happened here.

Within the first few months from the time I became a law enforcement officer, I realized the importance of absolute integrity in that type of job. I clearly knew and understood justice, what it meant, and how important it was to uphold. Another concept I also realized at an early age was karma. Karma refers to the spiritual principle of cause and effect. It refers to the intent and actions of people, which influence the future of the person that commits either good or bad actions. Karma is a cosmic law, a law that is above human law. No matter what you do on this planet, you will not escape from cosmic law in one form or another.

Sometimes it takes a long time for evil things people do to catch up to them, but I have found that sooner or later it does, and comes in many forms. Sometimes people leave this world never having the experience of the bad karma they have generated during their lifetimes. But, maybe they do. Perhaps something they have done that made a major bad impact on another person's life might trouble their conscience. And maybe they may feel a little sorry for what they have done to someone who didn't deserve what they suffered.

The murder case I will describe in the following pages has been, and will forever be, an example of the worst case of injustice I have ever experienced in my entire life. It has caused suffering for many people and families. It is full of lies told by the very people who we trusted to protect our communities and families. Justice was not served, but more important, testimony and facts were falsified. Evidence that would easily have exonerated an innocent person was concealed. An innocent person spent over 19 years in prison until his death.

When I was first asked by Jamie Baker, from Wallace, to write a history of Wallace, I had an inclination not to. However, I already had thoroughly studied Colonel Wallace's history and found that he had suffered many injustices. In my mind, this book would be karma for certain people. It gave me the chance to tell a story that will probably be the only bit of justice that will ever come about from what the reader is about to read. It is also my strong opinion that the evidence I acquired clearly supports my stand.

This was an interesting case, dogged by corruption, but most important, a case that needed to be exposed.

The Most Despairing and Corrupt Case I ever Worked and How the book *Breaking Blue* led to a crime committed involving some of the top law enforcement people in the state of Washington The Wallace connection – A predator in the brothels is protected by those responsible for protecting the public

I feel it is important for the public to know I have found, during my 26 years in law enforcement, that 90% of the police officers I have known and worked with have exceptionally high integrity. However, this was a major exception.

The type of conduct described in this section of this book is an embarrassment to the law enforcement profession – it does great harm to both the police and the public. The dishonest actions they took and the consequences of their actions has weighed heavy on me for over 27 years. This story needs to be told.

How it began

Following the May 15, 1987, murder of Gary Brigham in Pend Oreille County, it wasn't until June of 1990, that I got a break in my case. That break came in the form of Gene Barnes's brother-in-law, Herb Galbreath, who I learned had gone to the Bonner County sheriff's office and made a complaint that his brother-in-law, Gene Barnes had been killing people.

On June 5, 1990, Herb Galbreath came to my office in Newport and gave me a recorded statement about three murders Gene had bragged to him that he had committed. These included Gary Brigham in Pend Oreille County and David Richey in Spokane. Most notable, in his statement he told me things no one would have know, which matched the evidence of two crime scenes – the David Richey crime scene in Spokane and the Gary Brigham crime scene in Pend Oreille County.

Regarding the, *Spokesman-Review* paper carrier, murder of David Richey, he told how Barnes had described himself and another man beating Richey, and that Richey kept trying to crawl under the car his body was found next to. He described many more things about the Richey murder that only the murderer could have known. In regards to the murder in Pend Oreille County, he described who the victim was, who was with him, the degree of beating he inflicted on him, how he sodomized him with a wooden object, the same way that he had sodomized Richey, how he strangled him, how hard he was to kill, and the embankment where the body was thrown down.

Again on June 7, I asked both Gene Barnes's sister and brother-in-law to come to my office in Newport for the purpose of giving statements. Their statements were extremely enlightening to these murder cases.

The Bonner County contact

The witness to this story had been used before by the FBI and had provided solid evidence that had been used successfully. I then contacted Bonner County Prosecutor Phil Robinson and gave him a copy of the statements I had taken.

After my contact with Prosecutor Phil Robinson the following news clipping appeared in a news article following an arrest they had just made of Gene Barnes for the first degree murder of Vincent Lopez:

> *But the homicide investigations didn't come to full speed until early June when a fingerprint*

> *computer identified the body of Brigham in Pend Oreille County. "It's been a long time coming," Bonner County Prosecutor Phil Robinson said. "Our initial difficulty was the inability to identify our murder victim," Robinson said. "Until that identification was accomplished, there was absolutely nothing we could do."*
> *He credited Pend Oreille County investigators "with working very hard on this and doing some things that were beneficial to our case."*

This was an exceptionally easy case to solve. The Bonner County authorities did not make attempts to solve it as they were intimated by Gene Barnes. My office got involved during the time we were working our case, also involving Barnes. It is important to note that Robinson only had Barnes arrested after I told him I was going to contact the media and give them the information I had.

The Spokane contact with Chief Mangan

I then sent Chief Mangan all the statements I had taken with a letter explaining the circumstances. This did not go well. The fact that Spokane Deputy Prosecutor Clark Colwell, with Mangan's help, had just convicted a different person for that murder, they did not want this new information to surface no matter how strong this new evidence was. This is when they both began successful actions to keep this new evidence from ever being know during an appeal.

Prosecutor Tom Metzger of Pend Oreille County remains silent. He fails to even attempt to get involved in a murder in his county

Prior to being voted out of office, I prepared a 63 page affidavit for an arrest warrant, and my entire investigation for Gene Barnes for first degree murder and presented to Pend Oreille prosecutor Tom Metzger. He did not respond.

Some of the Statements Gene Barnes made to his sister and brother-in-law as a result of frequenting the brothels of Wallace

• "Me and Gene used to go cat housing a lot. We spent bucks at the cat house in Wallace. Eight, nine hundred dollars a night. We also had a mutual girlfriend. Me and the girlfriend used to talk when we were together."

"Gene does weird things to women." (Descriptions of weird and perverted sex acts described) (The main brothels they frequented at Wallace were the Lux and Luxette).

• "When we would go to the cat house, the broads would actually kick him out of the room. He would have a fist full of hundred dollars bills and they would send him walking."

• Later in interviewing both of Gene Barnes sisters they described numerous mean, perverted and sadistic things they witnessed Gene doing to animals and people. They described his temper as uncontrollable and violent, capable of harming anyone that crossed him.

The extreme dangers of prostitution

There is an important element to be aware of concerning the safety of brothels: When a prostitute works in a brothel she is surrounded by other women, also there are also other prostitutes, within earshot distance of each other. A brothel also has a madam who is always aware of what is going on. Consequently, there is safety in numbers. In Wallace, if there were problems with a customer, they would always call the police who would take care of the problem. The police in Wallace looked out for the prostitutes, as they should have. Just like the men working in the mines, theirs was a dangerous profession.

It is a well-know fact that many prostitutes are murdered or abused by a small number of their clients. These victims went willingly with their customers, leaving no word indicating who they went with or where they would be. As an example, from 1996 to 1998, Robert Yates murdered at least 13 women, all of whom were sex workers, working on Spokane's "Skid Row" on East Sprague Avenue.

What Spokane County Deputy Prosecutor Clark Colwell and Pend Oreille County Prosecutor Metzger Knew, but covered up

Between October 1986 and May of 1987, a seven month period, three men were murdered. Spokane Deputy Prosecutor Clark Colwell and Pend Oreille Prosecutor Tom Metzger knew all the details of these murders from the first to the last, and up until I left office in 1990.

• They were both aware, and had copies of all three autopsies, by Dr. George Lindholm, that tied them together, and the fact that two of my deputies attended one of the autopsies, where Dr. Lindholm advised my deputies to let me know of the identical ways our victim had been killed compared to Spokane's victim. They were also aware that Dr. Lindholm called me advising me of the major similarities in the Spokane and Pend Oreille County victims.

• They were both aware of the modus operandi of the suspect, Gene Barnes, and of the fact that he had been banned from all the brothels in Wallace, Idaho, because of his sadistic and perverted ways. Ways that were similar to the injuries inflicted on the murder victims, especially that both the Spokane and Pend Oreille victims had some object forced up their anus's hard enough to cause tearing to both. They both had their genitals injured, both were severely beaten and strangled, and Richey was nude from the waist down, and Brigham completely nude. They both had photos of the victims in the positions they were left when they were killed.

• Deputy Prosecutor Colwell was aware of the greasy substance that was all over the Spokane victim, David Richey's, hands, coat, and hair. Also, that police had placed brown paper bags on the Spokane victim's hands to preserve a greasy substance on his hands for evidence. Deputy Prosecutor Colwell also went through great efforts to keep this information out of the court system. He was also aware that this evidence never came out during the trial of the person arrested for the murder. Or that there were two witnesses that stated there were two people who were involved with the killing of the victim, David Richey, statements which he kept out of the court system.

Left: Spokane Deputy Prosecutor Clark Colwell Right: Pend Oreille Prosecutor Tom Metzger. *(Public domain)*

• Both Colwell and Metzger had the statements of Gene Barnes' brother-in-law, where Barnes bragged to him about all the killings, how and why he did them. He bragged that he and a friend of his killed David Richey, and during the process of this murder, how David Richey keep getting under the car to protect himself. (The reason for all the oily substance on his hands, hair, and body.)

• They were both aware of numerous facts the witnesses provided, which proved to be accurate, both in the Pend Oreille County murder and the Spokane murder.

• They were aware of when the pathologist, Dr. Lindholm, made a complete reversal of his findings concerning what he originally found in his autopsy reports. That Dr. Lindholm was now lying for the benefit of Spokane Police Chief Terrance Mangan, and Spokane Deputy Pros. Colwell.

• They were aware of the statements I took from the witnesses and the fact that they were used to convict Barnes in the first degree murder of Lopez in Bonner County. They were aware that I had over 25-years in law enforcement and knew how to take solid and accurate statements.

• They were aware of what Chief Mangan was doing and his motive for his lying. They were also aware of Chief Mangan's friendship with the Washington State Attorney General, Ken Eikenberry, and a number of his investigators.

• Prosecutor Metzger knew there was easily enough probable cause and evidence to charge Gene Barnes with first degree murder. He knew that – because I had presented him with a 63-page affidavit for an arrest warrant for Gene Barnes for first degree murder. He completely ignored that.

• They both knew I had contacted the top forensic consultant in the nation, Dr. William G. Eckert, who was the founder of Forensic Science Consultants International, and obtained a statement from him after I sent him the autopsies reports. Following his study of the autopsies, Dr. Eckert stated there are definite similarities in these cases.

Dr. Eckert then describes the many mistakes Dr. Lindholm made in the process of performing his autopsies. He bases this on the reports and several photographs I send him. He makes numerous corrections on how Dr. Lindholm's autopsies fail to clearly understand and present all the actual facts that were possible to gain, but were missed. Twelve years later Dr. Lindholm was charged with two felonies involving stealing drugs from dead people.

Both Pend Oreille Prosecutor Metzger, and Deputy Prosecutor Clark Colwell, like Chief Mangan, knew all the facts of this case. He knew all the players and exactly what they were doing. They knew all of the facts because I personally provided them with every bit of evidence as it came in.

During the process of Chief Mangan's efforts to protect Barnes and convict an innocent person, Prosecutor Metzger and Deputy Prosecutor Clark Colwell, both had the knowledge to come forward with the truth. Metzger chose to remain silent, and Colwell became part of the cover up.

• In a letter to the editor following Rowley's conviction Rowley wrote:

I have been following your articles about my case in your newspaper. I would you to keep a couple of things in mind. First of all, I feel real bad for the Richey family and send them my condolences. But on the other hand I am innocent of these charges regardless of what the jury verdict was.

Think about that real hard Mr. Caldwell, [Colwell] when you are boasting about your conviction rate. What about your over zealousness to obtain a conviction? What about an innocent man being sent to prison? I remained silent throughout the process with the press. It's time for the people of Spokane to wake up and realize that I am not a killer.

What this means is that a real killer is loose amongst yourselves, and no matter how many years I spend behind bars on this unjust conviction, no one can change the fact that I didn't commit this crime. I never realized that this type of railroading could actually take place in the United States of America, until now. I am not giving up, and I would like to thank my family and friends for their support. ***I pray that one day the truth will come out.***

Greg A. Rowley

Deputy Prosecutor Clark Colwell's motive

He didn't want a case in which he was responsible for convicting a man for first degree murder to be overturned. In spite of all the evidence that would have easily overturned the conviction of Rowley, Colwell fought hard and successfully kept this evidence from ever being used.

Pend Oreille County Prosecutor Tom Metzger's motive

About two years before this incident I presented Prosecutor Metzger with a letter of no-confidence, signed by all my deputies. They felt he was not doing his job. I never made this public, but they were right to complain. Metzger took exception to this, and rather than come out in defense of the truth in this case, remained silent. However, even worse, he failed to hold Barnes accountable for murder.

The chronology of events and individuals involved in the alleged Gene Barnes Suspected serial murder case

Written for former Sheriff Tony Bamonte

by former *Spokesman-Review* Reporter, Bill Morlin

News stories written by Bill Morlin, a reporter for the *Spokesman-Review*. Morlin has been in the news business for almost 50 years, and has reported on the majority of the cases that appear to implicate Gene Barnes in numerous serial murders. Morlin also believes that Chief Mangan's actions were done out of vengeance – the reason he wrote this chronology

Bill Morlin is a former Spokesman-Review Investigative reporter, He worked for the Spokesman-Review from 1972-2009. He has been a working journalist for 50 years. He started with the Associated Press, where he worked from 1967-1972. Since 2009, he has continued to work as a free lance journalist for various media outlets including the New York Times and others. He currently is writing for a nationalist news blog.

• Sept. 1, 1986 – Vincent Lopez, a one-eyed laborer from California, with a glass eye, disappears from his home in Spirit Lake, Idaho, after being accused of stealing earthworms and breaking a window in an apartment managed by Eugene Barnes, also of Spirit Lake, in Bonner County.

• Oct. 11, 1986 – The partially decomposing body of a "John Doe" is discovered by a hunter in a wooded area about 300 feet off Lightning Creek Road, eight miles north of Hope, Idaho, in Bonner County. At the time it was found the pants had been pulled down below the knees.

In 1990, the population of Spirit Lake, Idaho, was only 836, but it took almost four years for the body to be later identified as that of Vincent Lopez. Statements from Barnes' relatives led to solving the murder.

One of the family members who identified Lopez was Jody Barnes, Eugene Barnes' son, who said he was with his father Gene Barnes at the time Lopez was murdered. Jody Barnes said his father drove up an isolated road with Jody and Lopez in the vehicle. When Gene came to an isolated area , he told Jody to stay in the vehicle. He then walked Lopez up the road and out of Jody's sight. Jody then stated he heard three gunshots and went up the road, where he discovered his father over Lopez's body with Lopez's pants pulled down. Jody later stated he just made that story up. However, the evidence of his statement matched the crime scene.

• May 15, 1987 – The beaten, nude body of another man is discovered by a motorist on a shale slope near No Name Lake, off the Bead Lake Road, 12 miles north of Newport, Wash., in Pend Oreille County. Relatives of Eugene Barnes later tell authorities that Barnes confessed to them that he was responsible for the murder, which remains unsolved. The body is later identified, through AFIS (Automated Fingerprint Identification Search) as that of Gary Chester Brigham, 31.

• Feb. 24, 1987 – David Richey, a Spokane newspaper carrier, is found dead, brutally beaten and partially undressed alongside a car parked in a lot near Rogers High School in northeast Spokane. Greg Rowley, who lived nearby and steadfastly denied involvement in the murder, eventually was arrested and convicted in 1989 of the Richey murder. He was sent to prison where he spent 19 years, until he died there, still maintaining his innocence. Barnes' relatives say Gene confessed to killing Richey.

There was a witness to this who stated he saw two men beating Richey. Also, Eugene Barnes' brother-in-law said Barnes had bragged about Barnes and West killing Richey. During his description of that killing, the witness said Richey kept trying to get under a nearby car for protection. The photos of Richey lying dead next to this vehicle show a tremendous amount of dirt and what appears to be oil or grease on his jacket and in his hair.

• July 6, 1987 – Seattle Police detective Robert Gebo, who is an FBI-trained criminal profiler, says he sees similarities between the murders of Richey, Brigham and Robert McDonald, 40, whose body was found in the Pasco Boat Basin on Feb. 24, 1986, exactly one year before Richey was killed. Spokane Police Chief Mangan called the Seattle police department and advised them not to talk to the sheriff of Pend Oreille County if he called asking for information from Gebo.

• June 7, 1988 – Kevin D. Kent, 28, of Seattle, is found beaten to death in a car parked in a dirt lot outside Inland Bumper Repair, 2302 E. Trent in Spokane, WA, on June 7, 1988. The murder remains unsolved. In a statement from Herb Galbreath, Galbreath stated Gene bragged to him he also killed that person.

• Dec. 30, 1988 – John W. Deetz, 38, a truck driver from Illinois, disappears along with his new semi-tractor trailer from the Broadway Truck Stop, 6606 E. Broadway Ave, in Spokane Valley. A presumed victim of foul play, his body is never found. Witnesses later implicate Eugene Barnes in the disappearance and theft of the tractor-trailer rig and its owner. Eugene's son, Jody, also said his dad hit Deetz in the head with a tire thumper, unhooked the trailer and drove the almost-new truck away.

• February 1989 – Gregory Rowley is convicted by a state court jury in Spokane of the February 1987 beating death of newspaper carrier David Richey. Rowley flatly denies any involvement in the Richey murder, but admits he stopped by the murder scene, close to his home, leaving his bloody fingerprint, while in a marijuana-induced stupor.

• April 1990 – Eugene Leroy Barnes is indicted in U.S. District Court in Spokane on federal mail fraud charges following an FBI investigation implicating him in a truck-theft "chop-shop" operation. In the course of that investigation, court documents say, the FBI learned about Barnes' suspected involvement in the murder of Vincent Lopez and the 1988 disappearance of truck driver John Deetz.

• 1990 – Less than two years after Deetz disappeared, Ralph Howard Benson – a truck driver and Barnes associate – is spotted driving Deetz' missing 1989 International semi-tractor near Davenport, Wash.

• June 5, 1990 – In his first of two voluntary statements given to Pend Oreille County investigators, Herb Galbreath says his brother-in-law, Eugene Barnes admitted involvement with Arnold West in the beating death of Brigham. Galbreath also says "Gene (Barnes) told me that he also did that Richey murder."

Galbreath also says Barnes confessed to killing a "guy beat to death at a chrome shop on Trent" in Spokane (Kevin D. Kent, June 7, 1988?) Galbreath talks about providing those details to Spokane County Sheriff's Office, but nothing came of it. "Do you know how discouraging this has been? My God, man, the guy is killing people and nobody gives a shit. That's just the way it comes across to me."

• June 7, 1990 – Herb Galbreath gives his second voluntary statement to Pend Oreille County investigators. He is accompanied by his wife, Julie Galbreath, who is Eugene Barnes' sister. Both Galbreaths implicate Barnes in the murders of Richey and Brigham. They also mention the "disappearance" in 1987 of Chet Harkins, about 30, who lived in Spirit Lake and drove trucks for Barnes.

• July 5, 1990 – Information from the Galbreaths leads to the positive fingerprint identification of the "John Doe," found four years earlier, in October 1986, in Bonner County, Idaho. The victim is identified as Vincent Lopez, who allegedly broke a window in a Spirit Lake, Idaho, apartment managed by Eugene Barnes.

• July 16, 1990 – Eugene Leroy Barnes is arrested in Spirit Lake, Idaho for first-degree murder, after authorities used information from his relatives to establish probable cause for his involvement in the 1986 murder of Lopez. At the time of Barnes' arrest, it is reported that "at least eight law enforcement agencies, including the FBI, have taken an interest in Barnes in the past two years." At the time, Pend Oreille County Sheriff Tony Bamonte publicly says "an investigation into Barnes' activities also has turned up new information in the February 1987 killing of newspaper carrier David Richey in Spokane."

Spokane police, however, maintain the new information is of little value and they say they are convinced Gregory Rowley, convicted by a jury in February 1989, killed the 32 year-year-old carrier.

Bamonte, meanwhile, said Monday he is interested in exploring the possibility that the Brigham murder in Pend Oreille County may be linked to other killings. "I do want to ask for the public's help on this, Bamonte said." "I have information that there may be a connection between the Richey murder in Spokane and these other murders in Pend Oreille and Bonner County, and possibly others."

Information supplied to the FBI by Spokane County Sheriff's detectives, implicating Barnes in a truck-theft "chop shop" operation, led to the filing of the mail fraud indictment against him in April, authorities said. Barnes son, Jody Barnes, 18 was arrested Monday afternoon in Spokane County on a warrant naming him as a material witness to the Lopez murder.

Eugene Barnes' relatives, including his sister, Julie Galbreath, of Spokane, said they told Bonner County authorities several months ago why they believe Barnes may have had a role in the disappearance of the man they only knew as Vince. But for unexplained reasons, relatives said they saw the investigation stagnant.

"I'm real mad over the way this was handled at first in Bonner County, but now I'm glad something is finally happening," Julie Galbreath, the suspect's sister, said.

Similar comments came from another sister, Chey Austin, also of Spokane, who said she wanted to praise the thorough work done by Pend Oreille County investigators and the FBI.

The suspect's relatives said they provided information about the Bonner County murder to FBI agents, who were asking questions about Barnes' involvement in the interstate trucks-theft operation.

But homicide investigations didn't come full speed until early June when the fingerprint computer identified the body of Brigham in Pend Oreille County.

"It's been a long time coming," Bonner County prosecutor Phil Robinson said. "Our initial difficulty was inability to identify our murder victim." Robinson said, "Until that identification was accomplished there was absolutely nothing we could do."

He credited Pend Oreille County investigators "with working very hard on this and doing some things that were beneficial to our case."

• July 17, 1990 – Pend Oreille County Sheriff Tony Bamonte, in a newspaper interview, publicly advances the notion that Barnes may be a serial killer linked to other killings and disappearances, including the murders Richey and Brigham.

• July 23, 1990 – Spokane Police detectives interview Eugene Barnes, who denies any knowledge of the Richey homicide. Another possible suspect, Arnold West, also denied any involvement and passed a polygraph. But Barnes refuses to take a polygraph, saying that "he would have to take one in every case that he was a suspect in, and he did not want to do that."

• July 26, 1990 – An Idaho judge issues a warrant charging Barnes with possession of stolen property related to the Aug. 27, 1983, theft of two Chevrolet pickup trucks from Appleway Chevrolet in Spokane Valley.

• Aug. 2, 1990 – Kootenai County jail inmate Michael Dane Smith tells authorities that, during a jailhouse chat, Barnes implicated himself in the May 1987 killing of Gary Brigham, whose beaten body was found near Bead Lake in Pend Oreille County, and the February 1987 beating death in Spokane of newspaper carrier David Richey.

•Aug. 2, 1990 – The two sisters of Eugene Barnes, Chey Austin and Julie Galbreath, stand in front of the Spokane Public Safety Building and say publicly that their brother killed David Richey, but Spokane police chief Terry Mangan calls their claims hogwash. Mangan attempts to discredit Pend Oreille County Sheriff Tony Bamonte who sees similarities in the various murders and faces re-election.

• Aug. 9, 1990 – Dr. William Eckert, a nationally known forensic pathologist, says there is a "definite similarity" among three murders in Eastern Washington and North Idaho in 1986 and 1987. He was referring to the murders of Gary Bingham, Vincent Lopez and David Ritchey. All three were severely beaten, partially undressed and anally attacked, possibly with a club or baseball bat. Dr. Eckert also stated the genitalia were injured on both Richey and Brigham. Penis injured, possible bite on Richey and scrotum injure; on Brigham purplish swelling of the penis. Dr. Eckert didn't provide information about the 1988 beating death murder in Spokane of Kevin Kent.

• Sept. 14, 1990 – Spokane Police Chief Terry Mangan calls a press conference to announce results of an investigation by the Washington State Attorney General, upholding the Spokane Police Department theory that only Greg Rowley could have been responsible for the Richey murder and denouncing Bamonte for advancing the serial killer theory involving Barnes.

The unprecedented press conference by the Spokane chief, criticizing the elected official in another county, comes four days before Bamonte narrowly loses his primary election bid for re-election as Pend Oreille County sheriff. Bamonte believes Mangan's actions are "pay-back" and vengeful because of Bamonte's investigation a year earlier that revealed the Spokane Police Department was uncooperative in Bamonte's successful effort to prove that a Spokane police officer was involved in the 1935 murder of the town marshal of Newport, Wash.

• 1991 – Investigators find Deetz' missing repainted tractor in a lot in Los Angeles owned by Ralph Howard Benson, a known associate of Eugene Barnes, who lives in an abandoned missile silo near Davenport, Wash. The discovery of the missing Deetz tractor prompts Spokane County investigators to get a warrant to search Benson's missile silo home in Lincoln County, Wash. They find the still-shiny new license plate from Deetz' tractor during the search of Benson's rural, underground home in Lincoln County, Wash.

• June 12, 2002 – Roger Erdman, a state fuel revenue auditor, is fatally shot at the missile-silo home in rural Lincoln County, north of Davenport, of suspect Ralph Howard Benson, an associate of Barnes. Only parts of his dismembered body are recovered.

• July 2, 2002 – Ralph Howard Benson, 62, is arrested and charged with first-degree murder in connection with Erdman's murder.

• Nov. 13, 2003 – Ralph Benson is convicted of first-degree murder by a jury in Lincoln County. Afterwards, investigators say they are suspicious of Benson, who traveled the country as a long-haul trucker, who may have been a serial killer. His connections with Eugene Barnes are never fully publicly explained.

• Sept. 22, 2004 – Ralph Howard Benson dies of natural causes in the Washington state prison where he was serving 32 years for the murder of state auditor Roger Erdman.

When Gene Barne's brother-in-law originally contacted me at my office in Newport, WA, it was for the purpose of giving a statement regarding Gene Barnes and several murders Barnes had bragged about that he had committed. He also bragged about the details of how he murdered them. Once he started talking about these murders, I quickly recognized he was telling me things only someone would know if they had been talking to the actual person who committed the murders.

As he progressed with his preview of what he was going to say, within a short period of time, he came to a part of his story where he stated both he and Barnes used to patronize the brothels in Wallace on a regular basis. He further stated that because of the weird and violent perversions Gene would attempt with the prostitutes, they all refused to have anything to do with him. At this time I stopped

asking him questions and advised him that if I took a statement from him, his wife, (he was married to Gene's sister), would probably find out that he had been frequenting the Wallace brothels, and that may present a problem for him. I advised him to go home and think about it for a week or so, and if he decided to give a statement then he should tell his wife.

Two days later he made an appointment to see me again. I asked what happened and he stated when he told his wife, she stated she has known that for a long time. She stated: "Whenever we go down town to Wallace and pass any of the prostitutes on the street and they know and call you by your first name, its hard not to figure that out. Shortly after, I took recorded statements from both the sister and brother-in-law. At the time Gene Barnes sister and brother-in-law lived in Smelterville.

Spokane, Wash., Tues., July 17, 1990. THE SPOKESMAN-REVIEW A5

Spirit Lake arrest culmination of long probe

By Bill Morlin
Staff writer

SPIRIT LAKE, Idaho — The predawn arrest Monday of Eugene Barnes ended law enforcement's version of trying to thread a needle in a dark room.

At least eight law enforcement agencies, including the FBI, have taken an interest in Barnes over the past two years.

The 49-year-old truck driver was arrested on a first-degree murder warrant accusing him of the 1986 shooting death of Vincent Lopez, 30, who authorities say may have been killed for stealing a handful of earthworms and breaking a window.

His body was identified 12 days ago and the murder investigation was revived after information was provided by the suspect's relatives.

The relatives initially offered the leads months ago, they claim, but the investigation was pursued piecemeal — without an extensive exchange of information between law enforcement officials in Washington and Idaho.

But cooperation clicked in and the eye of the needle became clearer six weeks ago when a fingerprint computer identified a Pend Oreille County murder victim killed in 1987.

Now, law enforcement officials say Barnes is a subject of interest in these cases:

■ The May 1987 beating death of Gary C. Brigham, 31, whose nude body was found near Bead Lake in Pend Oreille County.

■ The December 1988 disappearance in Spokane of an Illinois truck driver, John W. Deetz. He and his tractor truck have never been found.

An investigation into Barnes' activities also has turned up new information in the February 1987 killing of newspaper carrier David Ritchey in Spokane, said Pend Oreille County Sheriff Tony Bamonte.

Spokane police, however, maintain the new information is of little value, and they say they are convinced Gregory Rowley, convicted by a jury in February 1989, killed the 32-year-old carrier.

But Rowley's attorney, Maryann Moreno, is expected to use the new information in a third attempt for a new trial.

Bamonte, meanwhile, said Monday he is interested in exploring the possibility that the Brigham murder in Pend Oreille County may be linked to other killings.

"I do want to ask for the public's help on this," Bamonte said. "I have information that there may be a connection between the Ritchey murder in Spokane and these other murders in Pend Oreille and Bonner County, and possibly others."

Information supplied to the FBI last August by Spokane County sheriff's detectives, implicating Barnes in a truck-theft "chop-shop" operation, led to the filing of a mail-fraud indictment against him in April, authorities said.

Federal prosecutors said they hoped the filing of that charge "would stir the pot" of investigative theories involving Barnes.

Apparently it worked.

In April, about the time Barnes was under federal indictment, he was identified by his relatives as a suspect in the Bonner County murder. But the victim of the 4-year-old North Idaho murder remained unidentified.

Lopez finally was identified as the victim after Barnes' relatives read a news article about the Pend Oreille case and offered investigators their details about the Bonner County case.

Relatives claimed that Barnes had talked to them about both killings.

That information was pursued when a Pend Oreille County Sheriff's Department investigator and an FBI agent traveled to California in early July to interview past associates of Barnes and Lopez, who positively identified Lopez as the murder victim.

Lopez, officers concluded, was the one-eyed man known as "Vince," who had disappeared from Spirit Lake on a summer day in 1986. The laborer disappeared, authorities were told, after being accused of stealing earthworms and breaking a window in a Spirit Lake apartment house managed by Barnes. Investigators have located witnesses who maintain they saw Lopez leave Spirit Lake in Barnes' truck.

Barnes' son, Jody Barnes, 18, was arrested Monday afternoon in Spokane County on a warrant naming him as a material witness to the Lopez murder.

Eugene Barnes' relatives, including his sister, Julie Galbreath, of Spokane, said they told Bonner County authorities several months ago why they believed Barnes may have had a role in the disappearance of the man they knew only as "Vince."

But for unexplained reasons, relatives said they saw the investigation stagnate.

"I'm real mad over the way this was handled at first in Bonner County, but now I'm glad something is finally happening," the suspect's sister said.

Similar comments came from another sister, Chey Austin, also of Spokane, who said she wanted to "praise the thorough work" done by Pend Oreille County investigators and the FBI.

The suspect's relatives said they provided information about the Bonner County murder to FBI agents, who were asking questions about Barnes' involvement in the interstate truck-theft operation.

He was indicted for mail fraud after he allegedly participated in a conspiracy to steal a tractor for an 18-wheel rig. Its owner, Peter Gordon A. Brown, also was indicted for making a fraudulent insurance claim for his truck, allegedly stolen and dismantled by Barnes. The case is set for trial next month in U.S. District Court.

In the course of the federal investigation, documents say the FBI learned about Barnes' suspected involvement in the Bonner County homicide and his possible connection to the December 1988 disappearance of John Deetz.

Neither Deetz, 40, nor his tractor truck have been seen since Dec. 30, 1988, when he pulled into the Broadway Truck Stop in East Spokane for food and fuel. His trailer was left behind.

Deetz' disappearance was investigated by Spokane County Sheriff's Detective Mike Massong, who told the FBI late last year about Barnes' possible connection to the stolen truck "chop-shop" operation.

But the homicide investigations didn't come to full speed until early June when a fingerprint computer identified the body of Brigham in Pend Oreille County.

"It's been a long time coming," Bonner County Prosecutor Phil Robinson said.

"Our initial difficulty was the inability to identify our murder victim," Robinson said. "Until that identification was accomplished, there was absolutely nothing we could do."

He credited Pend Oreille County investigators "with working very hard on this and doing some things that were beneficial to our case."

"The ultimate benefit to them, of course, is it may help their case" in Pend Oreille County, he said.

J. Barnes
Arrested

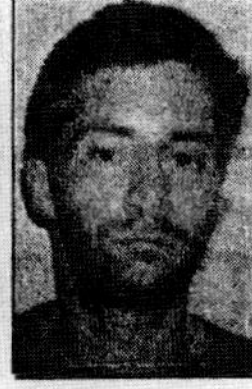
Brigham
Dead

Deetz
Missing

Lopez
Dead

A *Spokesman-Review* story written on July 17, 1990, talking about the arrest of Eugene Barnes.

The Gene Barnes Serial Murder Case And How It Was Covered Up By Public Officials

To me this was especially important, because it was an extremely, corruptly handled case by the former Spokane police chief, Terrance Mangan. **This case also involved Dr. George Lindholm the forensic pathologists who performed autopsies on all three victims, and later lied about the results. Dr. Lindholm was later accused and charged with two felonies. He was convicted of stealing prescription drugs from people whose bodies he examined in the morgue, and having a "grow operation" at his residence.**

• When Dr. Lindholm performed these autopsies he advised my deputy, and civil clerk, (both attended the autopsy) that whomever had committed the murder in Pend Oreille County was the same person that committed the murder in Spokane County. He called me the next day and advised me of the same thing.

• Later, they charged a person whose fingerprints they found in the car, and convicted him of first-degree murder. (Greg Rowley vs Richey)

Chief Mangan, with the help of Dr. George Lindholm, Washington State Atty. Gen. Kenneth Eikenberry, and two of his investigators, Bob Keppel and Bo Bolinger, and Spokane deputy prosecutor Clark Colwell, lied and completely denied any similarities in the deaths of all three victims. There was other extremely and crucial evidence, which, by them lying and spinning testimony, was never admitted.

This case began in 1986, and we worked it until I lost my election in 1990. The two deputies I had assigned to this, had spent time working on it. Besides myself, were my under sheriff, David Jackson, and my Sgt. Fred Warren. Both of these men will verify the corruptness of Spokane Police Chief Terrance Mangan, and the Washington State Atty. General and his investigators. Also, an investigative reporter (Bill Morlin), who was also reporting on this case and had been vetting the information prior to writing his articles, is also a witness to Chief Mangan, Dr. Lindholm, Deputy Prosecutor Colwell, attorney general investigators Bob Keppel and Bo Bolliger's false statements with deliberate attempts to mislead the public about the case.

• All of these people in high government positions of power, used their power to whitewash this case. Most important, their corrupt actions allowed a suspected serial murder suspect to go free. Strong evidence exists that this murderer may have committed at least four brutal and sadistic murders: Vince Lopez (He was charged for that murder, but as a result of Chief Mangan and his helper's actions the charge was dismissed). He was also a suspect in the murders of David Richey, Gary Brigham, John Deetz, and the Chrome shop murder in Spokane).

• During my 26-year career in law enforcement, this was the most corrupt police investigation I had ever witnessed, and it is the only case I ever copied and kept. This was fortunate, as there are no copies of it at the Pend Oreille County sheriffs office or at the prosecutor's office. I had also presented a 63 page affidavit for arrest for first degree murder, for the murder that had occurred in my jurisdiction. It also had disappeared.

Basis for Complaint:

The people named in the preceding page used the power of their high level public positions of trust, to publicly make false statements with deliberate intents to deny, distort, conceal, and assist in the exoneration of an extremely, dangerous serial murder suspect who had been active within a 100 mile range of Spokane. All of these identified public officials showed total disregard for the facts of this investigation and the safety of the general public.

Mangan's Motive

This subversive activity was accomplished during the course of an effort orchestrated by former Spokane Police Chief, Terrance Mangan. His motive

Terrance Mangan, former Spokane Police Chief.
(Public domain)

was a pay-back for exposing a 1935 murder case that involved the Spokane Police Department. In that case he was caught lying to a reporter working for the *Spokesman-Review*, Bill Morlin. Mangan's lie was publicly exposed in a book written by, Timothy Egan, a *New York Times* journalist and writer. Mr. Egan exposed Mangan lying, ridiculing, and calling press conferences using his influence and friendship with the above high-level public officials. Following being exposed by Egan, Chief Mangan's actions caused crucial evidence to be falsified and invalidated, and allowed a suspected serial murderer's first degree murder charges to be dismissed.

Falsification of facts relevant to three separate autopsy reports all performed by Dr. George Lindholm; These autopsy reports were completed on three separate dates. The first on 10-14-86, the second on 2-14-87, and the third on 5-15-87. All of these murders happened within a period of seven months. The murders that occurred on 2-14-87 and 5-15-87 had at least 7 extremely unique and sexually deviant similarities, according to pathologist Dr. Eckert. This falsification of evidence and cover-up was brought about with the approval and efforts of these same officials.

Case Synopsis

• On April 8th 1989, Gregory Rowley was sentenced to 46 years in the Washington State Penitentiary at Walla Walla, WA for the violent murder of David Richey.

• The evidence used to convict Rowley was circumstantial. There was no established motive or history of violence by Rowley. The Jury admittedly had a difficult time arriving at a decision.

• There is significant evidence and information which does implicate, beyond a doubt, another important suspect in the murder of David Richey. This other suspect is a convicted criminal, and does have a history of violence amenable to this type of specific crime.

• This information was provided to Chief Mangan of the Spokane Police Department by myself, and was ongoing on numerous occasions. Further, this information would be substantial enough to cause an appeal for Gregory Rowley.

• There was also other information and evidence which would have been in the best interest of Mr. Rowley that was discredited, some not obtained, and some concealed from the jury before and during the trial. That information consisted of statements from two other witnesses both stating two men killed David Richey. These statements were discredited and a witness, who overheard that conversation, was ignored (Callen).

• David Richey was murdered in Spokane on 02/24/87. Gary Brigham's body was found in Pend Oreille County on 5/19/87. Both of these men were murdered less than three months apart and within a radius of 50 miles from each other. **Both of these bodies had been assaulted in very unique ways and almost identical in some specific and sexual body areas of both Richey and Brigham. There were at least 6 common areas of injury. There were also other similarities. I provided this information to Chief Mangan on an ongoing basis.**

There was also a 3rd murder in Bonner County, approximately 80 miles from Spokane, which had similarities to both of the above murders and with a common suspect.

• Had this information been presented to the jury during Mr. Rowley's trial he would never have been convicted.

• During the last 45 days prior to my upcoming 4th election for Sheriff of Pend Oreille County, and at the request of Chief Mangan to his personal friend, the attorney general of Washington State, Kenneth Eikenberry, an investigation of my investigation was conducted by Bo Bollinger and Robert Keppel for the attorney generals office. On August 9, 1990, following a telephone request from Bo Bollinger, an investigator for the AG's office, I sent him copies of all three autopsies and copies of my complete investigation.

• Although this was a violent murder inquiry, with the high probability, based on all the evidence that was withheld from the jury, that an innocent man was in prison, **a time deadline was put on this investigation by the Attorney General. The deadline for this time limit was just prior to my election.** One week prior to this deadline I was contacted by Bo Bolinger. He asked if I could meet with him to go over the investigation. His phone call with this request was approximately 10:30 in the morning and he wanted to meet with me around 2:00 that afternoon. I told him I couldn't meet with him at that specific time but I could any other time. I further advised him I would travel to his location for a meeting if it would be more convenient for him. He informed me that was the only time he could meet. At this time, **I was informed by him, he had a deadline. His deadline was to enable a news story to come out just two days before my election.**

• The Attorney General's investigation was not conducted for the purpose of discovering the truth, or to determine if an innocent man was in prison, or if there was a dangerous murderer still on the loose.

His investigation was centered on discrediting my investigation. I was never given the opportunity to discuss this case with anyone from the A.G's office (other than the above 5 minute phone conversation). They had the entire case file I had provided, and whatever Mangan provided.

Most importantly they did not, at any time, contact the main witnesses: Chey Austin, Julie Galbreath, or Herb Galbreath, the two sisters and brother-in-law of the murder suspect, or the other two witnesses – one who stated he witnessed the murder and the other witness who witnessed that witness telling somebody about seeing David Richey in the process of being murdered by two people. That witnesses's name was Janette Callen. The Spokane police were using her house as a stakeout to catch a burglar who was a suspect for stealing things in that area. She was bent down behind her fence and working on her garden when the other witness came home excitedly telling his friend about seeing this murder. This witnesses name was Joseph Gariepy. He admitted to stealing a stereo out of a car, which he also confessed to during his statement. Gariepy's statement was discredited by the police and Callen's was never used. She was told by a police spokesman to forget it.

Jim DeFede, a reporter for the *Spokesman-Review*, knowingly wrote the false story concerning the suspected serial murder case. He also told me he had a deadline – with no time for my rebuttal.

• **According to Bollinger, because the A.G's investigation was on a time-line, it was necessary to be completed several weeks prior to my election. The results were given to Chief Mangan, who in turn immediately called a press conference inviting all the area media.** At this time, through the media, the above described people along with the Deputy Prosecutor of Spokane County, Clark Colwell, obstructed justice and grossly misled the public. The Attorney General's investigation was personally focused on falsifying all the facts that would easily have incriminated Barnes.

The main element that stands out in this case is the fact that Chief Mangan, the investigators for the Attorney General's office, Dr. Lindholm, the forensic pathologist that performed all three autopsies, Dr. Raey, and the Attorney General for Washington State, Kenneth Eikenberry, were all given copies of the autopsy reports that were done by Dr. Lindholm, and all of them consistently lied about the results.

It also included all of that evidence, and corroborating photographs in reports I sent them. Although there were two witnesses that testified to the fact that two people committed these murders, all of their testimony was discounted because they had criminal records. There was a third witness that had no criminal record, which was completely omitted.

As an experienced police investigator with over 25 years in law enforcement at the time, I knew for a fact, that the key to whether or not a jury will accept the testimony of a criminal is the extent to which the testimony is corroborated. All of the statements that were given by the witnesses could easily be corroborated. Also, the information in the statements they gave, could have only come from the knowledge of the murderer, or someone who saw the murders being committed.

A six page letter from Robert Keppel, investigator for the attorney general's office, was sent to Chief Mangan, the crux of Keppel's letter is extremely political and, rather than examining the facts, criticizes my interview statements with witnesses, especially Herb Galbreath's statements. This is significant because the majority of the information I received from witnesses could be proven to be correct. Galbreath had been used in the past as a credible informant for the FBI.

Robert Kepple, former investigator for the Washington State attorney general's office, knowingly falsified a report to assist Spokane Police Chief Terrance Mangan. *(Public domain)*

What is most irregular concerning Keppel's investigation is the fact that he takes simple informal tape-recorded interviews and attacks them as though they were sworn depositions, interrogatories, affidavits or actual sworn testimony under oath. Most important, he makes unsubstantiated and false claims concerning the substance of these statements with no evidence or chance for rebuttal ever offered or given. He then gives his findings to Chief Mangan to use in his press conferences.

When a police officer takes a statement from a witness there are three basic ways these statements are typically taken.

1) You can have the witness write out a statement in their own handwriting relaying what they are a witness to. These statements usually aren't very effective as most people need some type of guidance concerning what to write. They are often used when

there are many witnesses at a scene and limited police officers to record their statements. However, they are a good way of recording who the witnesses are and being able to record their addresses for later contacts.

2) The second type of statement typically taken by a police officer, is for the officer to question the witness and take notes during the interview. The officer later compiles his notes, at a later date, when the report is written. This often presents a problem as the witness does not get a chance to see what the officer attributes him/her to have said, unless the witness is given the chance to study the officers report, agree to it and sign that it is correct.

During the eight years I spent on the Spokane Police Department these are the types of witness statements that are typically taken. During my lengthy tenure in law enforcement I have experienced numerous occasions when statements have been attributed to witnesses and the witnesses have disagreed with what the officers have attributed to them. This was the way the statements were taken from the witnesses in the Richey murder.

3) The third type of statement taken from a witness by a police officer is the taped statement. There is a major difference between taking a taped statement from a suspect who is in grave danger of being placed in prison versus a friendly witness to a crime.

When taking a statement from a friendly witness the concept of the taped interview is to learn everything possible concerning the witnesses knowledge of the suspect and crime. These types of interviews are informal and often expand as the interview progresses. They also provide the most accurate record of what the witness knows and is able to testify to.

This is the type of statement I took from Galbreath. It was in a relaxed atmosphere, with Galbreath asking to stop the tape several times to ask his wife questions intended to refresh his memory. Anything said in his statement would have been subject to be sworn to under oath at a later date, or questioned by either a prosecutor, defense attorney or investigator for either side.

At the time of his statement I taped what he had to say to make an accurate preliminary record of what he claimed to know, which incidentally corroborated the evidence I had concerning Barnes. I learned this in 1966 at the SPD academy, and have done it, unchallenged, until Keppel's unfounded challenge in 1990. The most important element in taking a statement from a person with a criminal background is if it can be corroborated, which it easily was.

During Keppel's entire investigation I was never contacted by him or anyone from the AG's office. Why would a trained professional investigator dissect a statement and make it public without giving the originator or subject a rebuttal opportunity? **Why would there be a time limit to gather information on a murder case? It is also important to note that on the eve of my election Chief Mangan called the Pend Oreille County auditors office on five occasions, each time identifying himself and asking the results on the Sheriffs vote count.**

There is another important element concerning witnesses to crimes. Most witnesses to crimes don't have outstanding backgrounds; many have criminal records and behavioral problems; many witnesses have been in and out of, or are currently in jails or prisons when their statements are taken. Law enforcement doesn't have a choice of whom they can use as their witnesses. Numerous witnesses are often paid by law enforcement to be witnesses and many have their fines or sentences reduced. Being able to corroborate what they say is the key element.

Jim DeFede, former reporter for the *Spokesman-Review*, also had a deadline. He would not wait for the four hours I requested of him, and needed, to produce the autopsy reports, which would have clearly shown Spokane Deputy Clark Colwell and Dr. George Lindholm were lying.

Sisters say Barnes killed carrier, but Chief Mangan still disagrees

By Jim DeFede
Staff writer

The sisters of Eugene Barnes stood in front of the Spokane Public Safety Building Thursday and said they believe their brother killed newspaper carrier David Ritchey.

But the police refuse to investigate the case, the sisters said, because the detectives don't want to admit they helped convict the wrong man.

Thirty minutes later, Spokane Police Chief Terry Mangan defended his department, saying his detectives have reviewed "a volume" of material sent by Pend Oreille County Sheriff Tony Bamonte, but there is nothing to incriminate Barnes.

Mangan said his officers have taken "great pains to make sure the evidence is reviewed objectively," but he added, "we have found absolutely no new evidence.

"It's an old and tired story," Mangan said.

Mangan described the information as "smoke and mirrors" designed by Bamonte to garner media attention. It may make great reading, Mangan said, but none of it is valid.

Lt. Jim Hill, of the Spokane Police Department's homicide unit, said Barnes, who is charged with murder in Idaho and under investigation in at least one killing in Washington, is cooperating with Spokane police in providing information that may clear him in the Ritchey murder.

But Barnes' sisters said Thursday they would continue to speak out until the Spokane police realize they have made a mistake.

"The truth should be made public regarding the farce of this investigation," said Chey Austin, one of Barnes' sisters. "They made a mistake and they don't want to admit it."

"Nobody is going to shut us up," said Julie Galbreath, another sister.

Galbreath's husband, Herb, has been providing information to Bamonte, who is investigating Barnes for a murder in Pend Oreille County. Herb Galbreath has said that Barnes admitted killing Ritchey.

Mangan Thursday said Herb Galbreath's statements consist of "vague allegations that are contradictory."

Mangan and Bamonte both acknowledge that Galbreath suffers from brain damage caused by a truck accident, and has some memory loss.

But Bamonte defended Galbreath Thursday and said "everything he has told us is turning out to be true."

Please see **BARNES: B2**

CONTINUED: FROM B1

Barnes

For instance, Bamonte said Galbreath last week helped the Idaho State Police recover a stolen truck that Barnes had allegedly been using.

Mangan was still not convinced, and detailed for reporters the case his officers developed two years ago against Gregory Rowley, who was convicted by a jury of first-degree murder and sentenced to 46 years in prison for the Ritchey killing.

Mangan reminded reporters that Rowley's bloody fingerprints were found not only on the outside of the car where Ritchey's body was found but also inside the car, disputing the claim that Rowley just happened onto the body and inadvertently touched the car.

Mangan also said that Rowley repeatedly changed his story to police during their investigation when the evidence proved he was lying.

From the August 3, 1990, Spokesman-Review. *(Public domain)*

Staff photo by Chris Anderson

Bonner County deputies escort Eugene Barnes to court Monday for his arraignment on murder charges.

Spokane interested in Idaho suspect

By Greg Lee
Staff writer

■ Police try to link crimes – **A5**

SPIRIT LAKE, Idaho — Nearly two dozen state and local law enforcement agents surrounded a Spirit Lake home early Monday to arrest a truck driver accused of a 1986 homicide in Bonner County.

Following a three-day stakeout, authorities converged on the split-level home of Eugene Leroy Barnes, 49, at 4 a.m. and arrested Barnes without incident.

Bonner County sheriff's investigators allege that Barnes murdered Vincent Subia Lopez in a wooded area near Hope, Idaho, about Sept. 1, 1986.

Law enforcement officials say Barnes also is a subject of interest in the May 1987 beating death of Gary C. Brigham, 31, in Pend Oreille County, and the December 1988 disappearance in Spokane of an Illinois truck driver, John W. Deetz.

And Pend Oreille County Sheriff Tony Bamonte said he is trying to determine if there is a connection between the Brigham killing and the February 1987 murder of newspaper carrier David Ritchey, 32, in Spokane.

Please see **ARREST: A5**

The preceding page news article is continued below.

After Barnes was booked into Kootenai County Jail, he was taken in wrist shackles to 1st District Court at Sandpoint, where he was formally charged with first-degree murder. Magistrate Debra Heise ordered Barnes held without bail.

Bonner County Prosecutor Phil Robinson said Barnes will be held in Kootenai County Jail at least until his preliminary hearing, which must be held within two weeks. Robinson said that if Barnes is convicted he will seek the death penalty.

Investigators are sketchy about the date Lopez was killed because his body wasn't found until nearly two months after his death.

A hunter found the body about 300 feet off Lightning Creek Road, eight miles north of Hope, Bonner County sheriff's detective John Valdez said. The corpse was 40 percent decomposed, but an autopsy showed Lopez had died of three gunshot wounds to the head from a .22-caliber weapon.

Because of the body's condition and the absence of any identification, the body wasn't identified until earlier this month.

Lopez, 30, was a resident of Spirit Lake and had earlier lived in California, Valdez said. Lopez and Barnes were acquainted, he said.

"We have enough proof that he committed the murder," Valdez said. "We have a large number of witnesses that bear that out."

Valdez said detectives seized two firearms, a suitcase, shell casings and live ammunition during a search of Barnes' home at the corner of Seventh and Adams streets Monday. Valdez would not identify the type of firearms taken nor would he say

firearms taken nor would he say whether they were connected with the homicide.

Barnes' 18-year-old son, Jody Barnes, was arrested Monday afternoon in Spokane County on a warrant as a material witness to Lopez's homicide.

Jody Barnes' arrest is vital to the case because he has told several witnesses that he watched his dad shoot and kill Lopez four years ago, prosecutor Robinson told a judge Friday during the hearing.

Eugene Barnes also told a friend, Arnold West, that he killed Lopez, Robinson told the judge.

West's daughter, Virginia Swanson, was living with Lopez in an apartment building Barnes managed in Spirit Lake when Barnes ordered Lopez to leave town, West told investigators.

According to the accounts given by witnesses, Barnes picked up Lopez along the road as Lopez was leaving town. They drove to an undeveloped camping area northeast of Sandpoint where Barnes allegedly shot Lopez in the face, Robinson said.

A longtime neighbor of Barnes', Cynthia Neely, was surprised by Monday's pre-dawn arrest.

"I was shocked to death," said Neely, a retiree who has lived across the street from Barnes for 15 years. "I wouldn't think he'd be capable of murder."

CONTINUED: FROM A1

Inmate tells sheriff that Barnes discussed two other murders

By Bill Morlin
Staff writer

A Kootenai County Jail inmate's statements are being scrutinized by authorities who want to know if three murders committed in the region during 1986 and 1987 are linked.

Eugene Barnes, accused of murdering an itinerant laborer near Clark Fork, Idaho, in 1986, implicated himself in two other murders during a jailhouse chat in Coeur d'Alene with fellow inmate Michael Dane Smith, Pend Oreille County Sheriff Tony Bamonte said Wednesday.

The other cases are the unsolved May 1987 killing of Gary Brigham, whose beaten body was found near Bead Lake in Pend Oreille County, and the February 1987 beating death in Spokane of David Ritchey.

Gregory Rowley, whose bloody fingerprints were found on a car near Ritchey's body, was convicted in 1989 of murdering Ritchey, a 32-year-old newspaper carrier whose body was found near Roger's High School.

Smith, 21, offered new information about possible links between the killings Monday when he was called as a prosecution witness at a hearing in Sandpoint.

Bamonte attended Monday's hearing and later interviewed Smith by telephone.

In the interview, Smith said Barnes told him "that he had gotten other people (and) that he needs to cover up his trail if he ever gets out of here (the jail)."

However, Spokane police and Dr. George Lindholm, the forensic pathologist who performed autopsies on all three victims, have disputed the suggestion that the killings may be linked.

Eugene Barnes
TV spurred comments

"As far as we're concerned, it's a closed case," Spokane Police Deputy Chief Dick Jorgenson said of the Ritchey killing. "But we will, of course, look into new information that is supplied to us and provide it to the prosecutor."

After Bamonte informed Spokane police about Smith's statements, two detectives were sent to Coeur d'Alene on Tuesday to interview Smith, Jorgenson said.

"When our detectives interviewed him, he said Barnes made no statements relative to David Ritchey," Jorgenson said, adding detectives would like to listen to a tape recording Bamonte made of his interview with Smith.

Smith testified during the hearing Monday at which a judge ordered Barnes to stand trial on a charge of first-degree murder for the 1986 shooting death of Vincent Lopez in Bonner County.

The jail cell conversation between Barnes and Smith occurred after the two inmates watched a television news report of Barnes' arrest at his Spirit Lake home on July 16, Smith testified.

"He laughed like heck that nobody had figured out who had killed Brigham," Smith said in the tape-recorded interview with Bamonte. Bamonte played the tape for reporters.

When Barnes saw Rowley on the television news report, Barnes laughed again and responded: "That's the sucker that got nailed for this (Ritchey killing)?" Smith told Bamonte.

"He thought it was really, really funny that somebody else had gotten nailed for what he had done," Smith told Bamonte.

"He said he couldn't believe with the Ritchey killing that somebody else had gotten put away for it because he should have been put away," Smith said.

Smith added, "An exact quote he gave me is that he should have ran off

Please see BARNES: B2

Barnes

and left the state of Idaho two years ago because he knew the (expletive) was going to hit the fan."

Smith, who was serving time in the Kootenai County Jail for writing bad checks, first approached authorities after reading a news account that Barnes' son, Jody Barnes, had recanted statements implicating his father.

Smith said he wasn't given a break or lighter sentence for cooperating, and did so only because he feared an accused killer might be released, he said during the Monday hearing.

After testifying against Barnes, the two were returned to the Kootenai County Jail where, Smith said during the interview with Bamonte, Barnes "put his two hands together, kind of like you're holding a weapon, a handgun."

"It was like he was shooting at me, so it was real obvious how he feels about me," Smith told Bamonte.

Smith was extradited Wednesday to California, where he faces other check-fraud charges.

Barnes also is under federal indictment in Spokane, accused of mail fraud in a scheme that federal investigators say involved the theft and dismantling of a semi-tractor at the suspect's "chop-shop" in Spirit Lake.

An article that came out in the *Spokesman-Review* on 8-2-90.

Prosecutors blast Rowley's appeal

Evidence called 'questionable'

By Jim DeFede
Staff writer

Spokane County prosecutors filed more than 400 pages of legal arguments, police reports, sworn affidavits, newspaper clippings and medical records Friday in their answer to Gregory Rowley's appeal that his murder conviction be overturned.

An attorney is asking Superior Court Judge Thomas Merryman to set aside Rowley's conviction for the murder of newspaper carrier David Ritchey on the grounds there is evidence someone else may have killed Ritchey. Rowley is serving a 43 year sentence at the state penitentiary in Walla Walla.

A hearing on Rowley's motion is expected in a few weeks.

Rowley's attorney points to information provided by Pend Oreille County Sheriff Tony Bamonte that may implicate Eugene Barnes in Ritchey's 1987 beating death.

But Friday, Chief Deputy Prosecutor Clark Colwell described the new evidence as "unsteady, questionable, unreliable and unconnected hearsay" and asked the court to allow Rowley's conviction to stand.

The key in linking Barnes to the Ritchey case for Bamonte was Barnes' brother-in-law, Herb Galbreath, who said Barnes admitted killing Ritchey.

Galbreath, since making his statement has been largely discredited by Spokane police, based on inconsistent statements he has given to various law officers. They also cited his medical history, which shows he has both long- and short-term memory loss.

New information, released Friday by prosecutors, further impeaches Galbreath by revealing that in 1988 he was ruled mentally incompetent to stand trial on criminal charges in Yakima and that before then he has been hospitalized in Sacred Heart Medical Center's psychiatric unit.

Galbreath is a potential witness in Bonner County's case against Barnes for the Lopez killing.

Prosecutors also contend that Galbreath and his wife, Julie, were interested in getting paid to provide information against Barnes. And that Bamonte filled the Galbreaths' car with gasoline, bought the couple dinner and gave them $50 after one of their interviews with him.

Colwell, in his response filed Friday, also outlined the case that convicted Rowley, based in large part on Rowley's bloody palm print and fingerprints being found on both sides of the car where Ritchey's body was found. They also traced bloody footprints leading away from the body in a direct line to Rowley's house, across the street.

Rowley also was shown to have lied several times during his statements to police, changing his version of events every time new information against him was uncovered.

Through autopsy reports, Colwell also discounted the idea, promoted by Bamonte, that the Ritchey murder is connected to the killing of Chester Brigham in Pend Oreille County and of Vincent Lopez in Bonner County. Barnes is charged with the Lopez killing and is the leading suspect in the Brigham case.

"These alleged similarities are not supported by any medical testimony or exhibits, but only the opinions of Tony Bamonte, who is not a qualified expert," Colwell wrote.

Dr. George Lindholm, who performed the autopsies on the three victims, said the three cases are not related.

"It is unreasonable ... to suggest there is any relationship between the cases, in particular, a common party perpetrating all three crimes," Lindholm wrote. Ritchey had been beaten to death, Brigham had been strangled or hanged, and Lopez was shot in the head, Lindholm noted.

Bamonte had tried to dismiss Lindholm's comments by saying he was part of a cover-up by the Spokane Police Department to keep Rowley in prison.

But in addition to Lindholm's statements, prosecutors Friday presented letters from Donald Reay, the chief medical examiner for King County and the chairman of the Washington State Death Investigation Council, and William J. Brady, Portland pathologist.

After reviewing all three autopsies they also agreed that the three killings were not related and they complimented Lindholm's thoroughness.

Through autopsy reports, Colwell also discounted the idea, promoted by Bamonte, that the Ritchey murder is connected to the killing of Chester Brigham in Pend Oreille County and of Vincent Lopez in Bonner County. Barnes is charged with the Lopez killing and is the leading suspect in the Brigham case.

"These alleged similarities are not supported by any medical testimony or exhibits, but only the opinions of

Dr. George Lindholm, who performed the autopsies on the three victims, said the three cases are not related.

"It is unreasonable ... to suggest there is any relationship between the cases, in particular, a common party perpetrating all three crimes," Lindholm wrote. Ritchey had been beaten to death, Brigham had been strangled or hanged, and Lopez was shot in the head, Lindholm noted.

This article appeared in the Sept. 22, 1990, *Spokesman-Review*: It was important for Deputy Prosecutor Colwell to conceal the actual evidence and not let it make it to the appeal process. Consequently, to keep his verdict of first degree murder valid, he needed to breach his integrity and lie about something that was easily provable to be a lie. The people that did this used the power of their positions to be able to accomplish this deception to the public. When the reporter, DeFede, made a surprise visit to my office to interview me, my files were at my residence, 50 miles away. I told him if he give me the time to retrieve the autopsy reports they would prove what he was going to print wasn't the truth. He stated he had a deadline on his story and didn't have time. The most significant thing about this story is that it came out two days before my election. During the eve of my election, Chief Mangan called the Pend Oreille County Auditors office five times, each time specifically asking for a vote count on the sheriff's election. During these calls he made it a point to identify himself to P.O. Auditor Ann Swenson.

Following the below letter to Stacey Cowles, Publisher of the *Spokesman-Review*, I received this message back from him on May 5, 2017:

> *Thank you for sending me your extensive file. I've forwarded it to our Managing Editor Joe Palmquist; I believe he'll have Tom Clouse review it and get back to you. In my brief scan and from what you sent me before, I can't dismiss your claims. . . .* [I appreciated Mr. Cowles response]

On May 8, 2017, I received an e-mail from Tom Clouse asking for a contact number. He called the next day and stated he was going to look into this. Twenty-seven years ago, immediately following my conversation with Mr. DeFede, I attempted to contact Bill Cowles with no response. I have tried on numerous other occasions to bring justice to this wrong, with no success. Justice has still not been served.

April 30, 2017

Stacey Cowles, publisher
Spokesman-Review
999 W. Riverside Ave.
Spokane, WA 99201

Dear Mr. Cowles:

My reason for this letter is to be able to state, and clearly show, that I have given the top people at the *Spokesman-Review* ample evidence to prove that articles Jim DeFede wrote regarding the David Richey murder case in the 1990s were not truthful and did harm to many people. I have also tried on other occasions to get your paper to correct what was written about this case that let an innocent man spend 19 years in prison, until he died there, and to let an exceptionally dangerous person remain free.

I am enclosing a 291-page report. This report includes autopsy reports, photographs and statements that completely contradict and prove that former Spokane Police Chief Terrance Mangan; former pathologist employed by the city of Spokane, Dr. George Lindholm; former Spokane Deputy Prosecutor Clark Colwell; former Washington State Attorney General Ken Eikenberry; and former investigators for the attorney general, Robert Keppel and Bo Bollinger, were all lying to conceal and change evidence in a murder case, and did in fact change this evidence to conceal the truth in a first degree murder conviction.

I respectfully ask that you study the materials I am providing and, after 27 years, print the truth. I ask that you start with the enclosed September 22, 1990, article by Jim DeFede. In that article Deputy Prosecutor Colwell makes the statement: "These alleged similarities are not connected by any medical testimony or exhibits, . . ." Then look at Lindholm's actual autopsy reports. Dr. George Lindholm then makes the statement: "It is unreasonable to suggest there is any relationship between the cases, in particular, a common party perpetrating all three crimes, Ritchey had been beaten to death, Brigham had been strangled or hanged, and Lopez was shot in the head," Lindholm wrote. See Lindholm's autopsy reports for the many sadistic, bizarre, and highly similar injuries to all the victims. It is also important to note that 12 years after Dr. Lindholm had lied regarding this case, he was charged with two felonies for stealing prescription drugs for dead people.

Please know that my intentions are to seek justice in this case. This morning I read an article in the S-R by Thomas Clouse. Regarding the thing that stood out the most about this article (See enclosed) was Prosecutor Haskel statement: "In any murder case, the medical examiner is one of your star witnesses, . . . your medical examiner is going to provide you a significant portion of those elements of the crime." I ask that you turn this letter over to Mr. Clouse, after you have studied it, to show him the damage that can be done by a pathologist, which is similar to what is happening in the case he wrote about. The main difference in the cases is Dr. Lindhom blatantly lied, as he knew Chief Mangan and Deputy Prosecutor Colwell wanted him to do.

Most important, this information is going to be made public. This is the last, and most thorough amount of information I will have provided the S-R to this date. I would be pleased to met with you at your request and go over this case. If I do not hear from you within one week I will understand that you are not going to respond.

Sincerely,

Tony Bamonte

Chapter XI

A Family History My Father, Louis Bamonte, And His Family – Dale, Tony, and Star

The Biggest Influence in My Life, My Father

My dad, Louis Bamonte, was born in New York City in 1906 to Grazia "Grace" and Domenico Bamonte. They came from Roccadaspide, Italy, a town in the province of Salerno in the Campania region of south-western Italy. His father passed away in 1910, and three years later his mother passed away, leaving their six children orphans. At that time, my dad's older brother and sister, Ralph and Edith, took over raising the family.

In 1926, at the age of 19, my dad hitch-hiked across the United States to San Francisco, and joined the Marine Corps. Most of his tenure in the marines was served aboard the *USS Maryland* and in Nicaragua. The United States Marines occupied Nicaragua from 1912 to 1933, except for a nine-month period beginning in 1925 with the evacuation of U.S. Marines. Another violent conflict between liberals and conservatives took place in 1926, which resulted in the return of U.S. Marines. It was during this time that my dad was stationed in Nicaragua. From 1927 until 1933, there was a sustained guerrilla war against the Conservative regime and subsequently against the U.S. Marines. My dad was in Nicaragua from 1926 to 1928.

Louis Bamonte, United States Marine Corps from 1926 to 1930

In 1930, following his discharge, my dad had become friends with another marine from Kellogg, Idaho, George Harvey. As the United States was

Louis Bamonte on far left. One of the many mining leases my father was involved in during the time he spent in the Coeur d'Aenes. This photo of him and three partners was at the portal of the mine. This mine, including many of the smaller mines had no water or air to run any types of drills. Consequently, the blasting holes were all drilled with hand steel. This is the same type of mining August Paulsen was doing when he discovered the Mighty Hercules Vein. *(Bamonte photo)*

This is the site of the Oriole Mine located at the base of Mt. Linton in Pend Oreille County. It was my dad's last mining venture.

The Oriole was a lead, silver and zinc mine. The arrow indicates the portal to the mine. Pictured is the ore bin to the far right, the compressor shed in the middle, the blacksmith shop to the left, a partial of a small cabin to the far left, and to the lower right is a horse garage.

When my father died, he asked that I scatter his ashes near the portal of this site, which I did. *(Bamonte photo collection)* just in the beginning stages of the Great Depression, employment was scarce. However, there was lots of work in the Idaho mines, which was my dad's next move. From the age of 23, he worked for over 18 years in the various mines, and logging, in the Coeur d'Alenes – mostly the mines. Today, as I think back, I wished I would have asked him more questions about his early life. I know of three mines where he worked: the Bunker Hill, Sunshine, and Polaris. I know he lived in Wardner when he first came to Shoshone County. He lived there for several years until his house burned down. He then moved in, as a boarder, with the Gordon family in Wardner. During that time he became friends with Joe Gordon, who at one time became the general mine manager for Bunker Hill. Joe was several years younger than my dad, but they formed a lasting friendship, I'm sure, based on mutual respect.

Joe Gordon
My second most important influence

In 2004, a book came out, written by Fritz Wolf who spent time working at the Bunker Hill. Wolf spent ten years in the mining business and twenty-three in aerospace management. The title of his well-written book was, *A Room For the Summer, Adventure, Misadventure, and Seduction in the Mines of the Coeur d'Alenes*. His book is 264 pages in length. One of his chapters is titled "Joe Gordon." He begins his chapter describing Joe Gordon:

> *Joe Gordon was a mustang, working his way through the herd to become manager of mines. That position was the top of the heap in terms of mine operations. He oversaw all the Bunker Hill production, as well as leases outside the valley, like the Nancy Lee in Montana and the Sullivan near Metaline Falls, Washington. Joe was a local boy. He graduated from Kellogg High School in 1930 and worked underground for ten years to help support five siblings. He put himself through the University of Idaho's mining engineering program and returned to the Silver Valley. Joe started in management as a shift boss, then division foreman, mine superintendent, and finally, by the time I arrived, manager of mines. He knew the Bunker Hill upside down and backward. Joe took mining seriously. When it came to taking care of the men, he said what he meant and meant what he said.*

Betty and Joe Gordon at their Wedding ceremony, 1943. *(Courtesy Janice Gordon, daughter)*

I should also mention that Joe had a profound effect on my life. When my dad passed away in 1961, Joe and his wife were always there for me, and as best they could, filled the void that my father left with his passing. With them in mind, I would like to record his memory in this book. The following was from Joe's obituary:

Joseph E. Gordon, 94, residing in Coeur d'Alene; died Sunday, March 9, 2008. Joe was known for his intelligence and integrity, his strong work-ethic, his loyalty to and love of his family, his sardonic and quick wit, and his love of mining. He spent his entire professional career with Bunker Hill Company in Kellogg. He began working there when he was 17, and after having earned his way through the University of Idaho, obtained a B.S. degree in Mining Engineering. He also attended the Colorado School of Mines and attended the Advanced Management Program at Harvard University. He became a professional Mining Engineer and Geologist, and eventually became Manager of Mines at Bunker Hill and Senior Exploration Engineer.

Joe was born in Kellogg Feb. 23, 1914, to Catherine (Maguire) and George Gordon, both of whom had emigrated from Ireland. He grew up in Wardner and Kellogg and returned to live there after he attended the University of Idaho.

In 1943, he married Elizabeth Hall of Pocatello, who also graduated from the University of Idaho. They had two children, John and Janice.

Attaining the rank of Captain in the U.S. Army Infantry, Joe served in both World War II and the Korean War. He commanded 200 paratroopers, and was later appointed as a General's aide at Fort Benning, Ga.

His first wife, Betty, died in 1980. He married Margaret (Caruthers) Ritzheimer in 1981, and resided in Lake Oswego and Eugene, Ore., until her death. In 2006, he moved back to Coeur d'Alene to live with his son and daughter-in-law, John and Vicki Gordon.

Joe served on the Idaho Commission on Federal Land Laws, was a member of the Legion of Honor in the Society for Mining, Metallurgy and Exploration; a member of the Society of the American Institute of Mining, Metallurgical, and Petroleum Engineers; and a member of the International Precious Metals Institute Inc.

The Hazards of the Mines

In the history of the mining world, and specifically the Coeur d'Alenes, mining has always been a dangerous and unhealthy way to make a living. Dur-

The Joe Gordon family. From left: John, Janice, Joe, and Betty. *(Courtesy Janice Gordon, daughter)*

The mining and logging town of Metaline Falls.
(Bamonte photo)

ing my lifetime, I have spent a little over two years working in the Pend Oreille mine. Before I quit I was working as an apprentice miner. Prior to that I was beating boulders through a grizzly, mucking belts and driving diesel trucks underground. The most important thing I learned was that mining was not a career my father wanted me to follow. In 1963, at the age of 61, my father died of "Miners Consumption." This was a disease common to anyone that worked in the mines. It's caused by the inhalation of particles of industrial substances, particularly inorganic dusts, and silica. Symptoms include shortness of breath, chronic cough, and expectoration of mucus containing the offending particles.

In his well-researched book *From Hell to Heaven, Death Related Mining Accidents in North Idaho*, Gene Hyde documented with names and incidents, 3,238 mining deaths in the Coeur d'Alenes from 1887 to 2001. These are only the mining accidents. The graveyards in the Coeur d'Alenes are full of men, like my father, whose lives were shortened by breathing the dust and oil fumes of the mines.

Underground photo taken in the late 1880s of one of the mines in the Coeur d'Alenes . *(Butch Jacobson collection)*

Left photo: My brother, Dale Bamonte; my father, Louis Bamonte; myself, shortly after I was born on May 1st, 1942, at the Providence Hospital in Wallace, and my mother Lucille "Starmer" Bamonte. The photo on the right is my mother just before they were married. My Brother, was also born in Wallace in 1940. At the time of my mother's death she had been married eight times. *(Bamonte photo)*

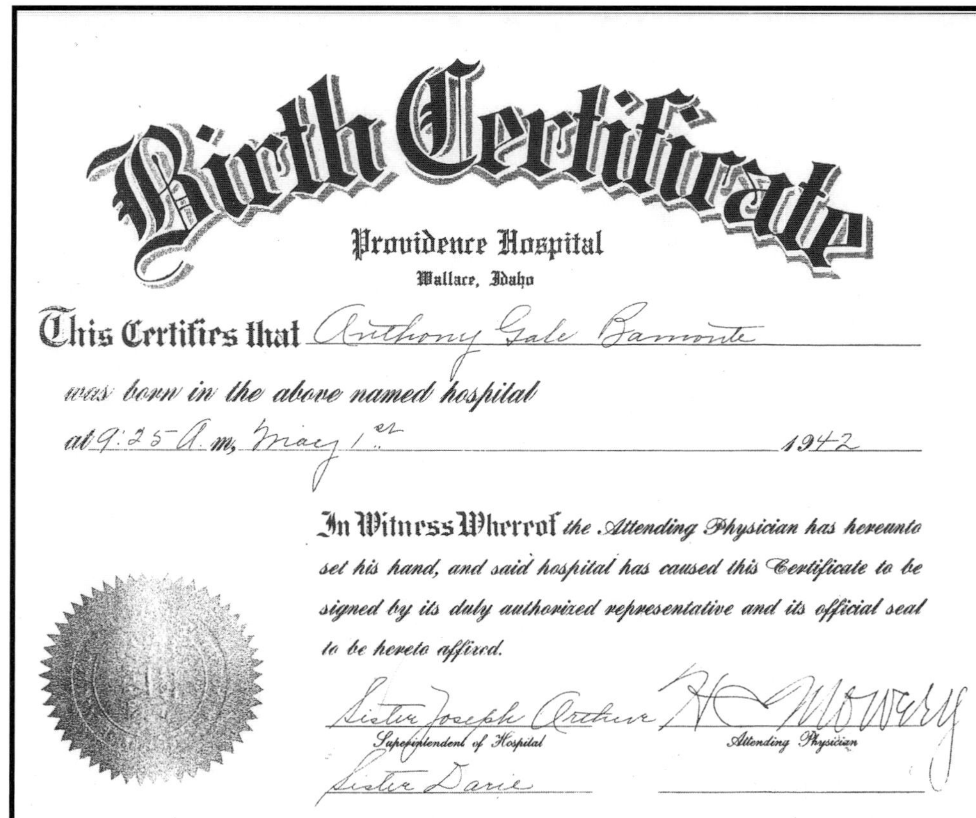

Birth Certificate

Providence Hospital

Wallace, Idaho

This Certifies that Anthony Gale Bamonte

was born in the above named hospital

at 9:25 A.m, May 1st 1942

In Witness Whereof the Attending Physician has hereunto set his hand, and said hospital has caused this Certificate to be signed by its duly authorized representative and its official seal to be hereto affixed.

Sister Joseph Arthur — Superintendent of Hospital

H. C. Mowery — Attending Physician

Sister Darie

My birth certificate from the Providence Hospital giving my date of birth on May 1, 1942. Placer Center (Wallace) was platted on May 1, 1884.

My birth certificate gives my middle name as "Gale," as I was named after my mother's brother. In 1950, Gale and my father had an altercation at one time. I was there at the time and saw my father easily get the best of him. Shortly after, I was baptized as a Catholic. When I was baptized my dad changed my middle name to George. We both liked it better. *(Bamonte photo)*

The move to the Metalines area

By 1946, mining had left my father in poor health and susceptible to frequent bouts of pneumonia, and as my mother didn't want to live that far away from a large city, he moved his family to Spokane. At the time, my brother was six, my sister wasn't born yet, and I was four. In the winter of 1949, while my dad was working in the Metalines, my mother started having a relationship with another man. Before long, with my father working in the mines and woods in Pend Oreille County, my mother began an affair. On December 25, 1948, on Christmas morning my father confronted both her and her lover. When the police arrived my father had beaten the man unconscious, with the intent of beating him to death. Shortly after, she left her family and moved to Hawaii. For two months my brother, sister and I lived with my maternal grandparents. At the end of the two months, and shortly after midnight, my grandparents got into an argument. My grandmother wanted to lie in court to help my mother during her trial for child abuse, and my grandfather was adamant about telling the truth. In the middle of the argument my grandfather suffered a stroke and died in their home.

Living next door to a former brothel

My father's next move was to relocate his kids to an old dance hall building, ten miles north of Metaline Falls and two miles south of the Canadian Border.

The "Old Red Rooster," where my family lived for over seven years. The horse we are on was named King. He was a kind and gentle work horse that was shot mistakenly by a poacher. When my dad found him dead it was a heart breaking event for our family. The cabin on the right is where my family lived. The partial log house on the left is part of what was at one time called a crib (a small duplex brothel). On the horse, from left to right is my brother Dale, sister Star and myself. *(Bamonte photo)*

My sister Star, and my aunts. *(Bamonte)*

From left: Star, Dale and Tony. *(Bamonte)*

My brother Dale, my dad and Sister. *(Bamonte)*

My sister, Star, holding her first born, Lani Jo. Star was 15 1/2 at the time, and had run away from our home, as our dad was strict. Her daughter, Lani, turned out to be beautiful – both inside and out, and a blessing for the family. *(Bamonte)*

Living at that location allowed him to work in the woods in the summer making cedar post, and later in the cedar pole business and logging.

The house we moved into was formerly known as the "Old Red Rooster," an early 1920s dance hall. Several other buildings surrounded it – a small log "crib" duplex and a 15' x 15' frame cabin. There was no indoor water or plumbing in any of the buildings, one light bulb, no telephone, and wood for heat. We lived there six years. During the summers, my dad worked in the woods as a cedar post maker. To keep from being left alone during the summers, we went to work with him. During that time, I became proficient with splitting, peeling, and trimming fence posts. In the meantime, my dad made arrangements for my sister to live with a family in Metaline Falls, with the exception of week ends. Since we had no indoor plumbing he felt that was the best for a four-year-old girl.

When my parents divorced, it was an easy decision for me to be with my dad. He was the most important person in the world to me. My dad was totally devoted to raising his children. He never remarried, drank or smoked, always worked hard and was exceptionally honest. The only entertainment he ever indulged in was reading. On Saturday evenings he would take us kids to the movies at Metaline Falls. His life consisted mainly of working by himself in the woods, year after year, with the only exception being when I was with him, which was as often as I

The "Old Red Rooster" again. Directly behind our horse was the original dance hall where we first lived. The two cribs were on the right. When we first moved there, another couple lived in the house to the right. Occasionally, my dad would have his wife baby sit my sister when he was working. Her husband owned a shingle mill about a half mile away. One evening my dad came home and found her husband had molested my little sister. Immediately behind the cabin was a water trough. I heard some noise and went to investigate. My father had the man begging for his life, swearing he hadn't touched my sister. Two days later, the man's shingle mill burned to the ground and they left the area. I'm almost sure he burned it himself out of fear for my dad. Several weeks later we moved into the abandoned cabin as it was a much better place to live. *(Bamonte photo)*

could. I worried constantly, especially since I knew he had health problems – the result of working in the mines for so many years. I often think how lonely and hard his life was.

Among the best times in his life and when I saw him the happiest was when his three sisters travelled from New York City to spend over a month with us. In 1954, they traveled by train from New York City to Spokane, where my dad and kids picked them up. From there we drove to our home near the Canadian

This photo is of my dad's 1940 Chevrolet truck with a load of fence posts. From left: My dad's middle sister, Rose, my dad, his youngest sister, Ann, his oldest sister, Edith, my brother, Dale, my sister Star, and myself. *(Bamonte photo)*

border. That was the first time any of us kids had ever heard our father speak Italian. It was also one of the best times of my life.

The worst day in my life involved my dad, in the fall of 1951, when I was nine, my brother and I had just gotten home from school, a 10-mile school bus trip from Metaline Falls. Typically, when my dad was working in the woods, he would get home sometime between 6:00 and 7:00. By 9:00 P.M. he still hadn't come home, and it was completely dark out. We had no phone and lived two-miles from the Canadian Border, which shut down at 6 P.M., meaning there was no traffic on the road. Consequently, there was no one I could ask for help. I needed to find him. In my mind, I was sure he was injured or even dead. I tried to get my brother to come with me, but he felt he needed to stay at the house if anyone did, by chance, come by. I knew the immediate area where my dad had been working, as we worked with him on weekends. It was about five miles from our house.

That night it was exceptionally dark; the moon was about at a quarter of its glow. I had a flashlight to use for going to the outhouse after dark, but the batteries were almost dead. As I started out I used the flashlight as little as possible, only enough to see where I was walking. I felt it was important to save the light for when I found my dad. All the time I was walking my mind was focused on my dad and why he didn't come home. This had never happened before, he was always there for us. My thoughts were of finding him dead or injured. He had an old 1940s truck he always drove. I watched him drive and felt confident I would be able to drive it if I went slow. I had studied what he did when he drove. I knew I would only have to use one gear to make it back home.

As I continued my walk, I sang what songs I knew. This was my attempt to scare any predatory animals off, as we had seen both a cougar and a wolf recently. My biggest fear was finding my dad dead. I walked in almost complete darkness for about three miles, as my mind raced to the various scenarios

The Bamonte family getting in wood. The house on the far right was the former dance hall. The washing machine sitting on the porch had a gas motor to run it. However, it didn't work. My Dad had a Kodak camera that had a built-in timer to allow for the entire family to be in the photo. The chain saw my Dad was using was an IEL, which was his first gas-powered saw. *(Bamonte photos)*

I might encounter. Suddenly I heard the noise of a truck moving toward me from about a half mile away. It was my dad. His truck had been stuck in a cedar swamp near where he was working. This was the worst/best day of my life.

The Move back to Osburn, Idaho

In 1955, my dad got a timbering job at the Polaris Mine. It was to last about a year. During the years he had worked in the Shoshone County mines he had acquired a reputation of being a good worker and dependable. We packed up all our belongings on my dad's 40 Chevrolet truck and moved to Osburn, where, for a year we rented a two bedroom unit at a place that was called the Royal Autel, which has since been torn down.

My most memorable experience at Osburn was when my dad made me take accordion lessons. He found out there was a man in Wallace giving accordion lessons, bought an accordion and enrolled me for lessons. I remember three things about this experience: 1) My accordion teacher's name was Mr. Arnold. 2) Mr. Arnold formed an accordion band. There were ten kids in it. He would have ten chairs placed all in a single row facing an audience of about 20 people and we would all play accordion songs, and 3) I wasn't good.

By 1956, my dad's contract at the Polaris had been completed. In the meantime he got an offer to go back to Pend Oreille County and construct a small concentrator mill at the first mine to ever produce and ship ore in the county, the Oriole Mine. We then moved to the town of Metaline Falls, in the same 1940 Chevrolet truck my dad had owned for years.

Prior to beginning construction of the concentrator mill, the people who owned the mine wanted him to recruit other men to begin developing the mine. One of those he recruited was Bill McCoy from Kellogg. Bill and his family moved to a small cabin next to ours on the mine site. Bill's family consisted of his wife, Sally, and their three children, Terry, Ronnie, and Don Marie.

They were at the mine site for almost ten months. I became friends with their second son, Ronnie, who was my age. His ambition in life was to become a jet fighter pilot. When they moved, I lost track of the family for about ten years. I later learned he made it to the Air Force Military Academy and was training to fly fighter jets. He died in an in-air flight accident with another jet fighter pilot trainee when they collided. Both pilots died instantly. Also, the Oriole Mine venture turned out to be a flop.

My dad, Louis Bamonte, in the spring of 1956, building a concentrator mill at the Oriole Mine (a silver and lead mine) west of Metaline, Washington. The Oriole was the first producing mine in Pend Oreille County. My dad was in partnership with four other men. When my dad passed away, his last wish was to have his remains scattered at the mine site. I scattered them 60-feet northwest from where he is standing in this photo. *(Bamonte photo)*

Our family lived at the mine site until fall when we moved into a small house in Metaline Falls. My dad then again went into the cedar pole and logging business. From 1957 to 1961, until I graduated from high school, I worked for him peeling cedar poles at $5 a day. The cedar-pole-peeling business begins in the spring, when the bark starts slipping and ends in the mid-summer when the bark starts to stick.

Things that stand out in my mind: Poles should always be peeled on the same day the trees are sawed down or the bark will begin sticking. The average length of the poles we peeled was from 25' to 65'. We peeled a lot of 45 to 50 foot poles. When you peel a pole, you peel up one side and then down the other. After the first bark strip is taken off both sides, it becomes easier to peel. I was able to peel from six to twelve poles a day depending on the diameter and length of the pole. Also, if I recall correctly, B.J. Carney and Company paid approximately $20 for a 35' cedar pole and $40 for a 65' pole, with the payment increments adjusted accordingly for the sizes of the others.

I spent many hours after school, on weekends and during summer vacation working with my dad. He passed away in 1966 at the age of 61, from Silicosis, the result of his years in the mines. Not a day goes by that I don't think about him and his influence on my life – mostly about the tough life he lived.

In 2016, two friends and myself began working on a book about police motorcycle officers, city, county and State Patrol, a job we all did while on the various police departments. Jack Pearson rode motor for the Spokane Police Department in the 1970s, Tim Downing, rode motor for the Spokane County Sheriff's Department, in the 1970s and 80s. Tim also taught motor school. I have great respect for both. In 2017, Tim's father passed away. I never personally knew his father. However, I attended his funeral out of respect for Tim. His mother asked him to speak at the service. What he said struck me as what I wish I would have said when my father died. I was much younger then and emotions and maturity worked against me. The words Tim spoke about his father made it the most touching funeral I have ever attended. The following is a copy of his speech – the same speech I wish I would have given for my father:

The cabin on the left was our home from 1956, up until my dad died in 1963. My beloved dog, "Butch," is in the center of the photo. The vehicle on the right was a 1942 Studebaker Champion, which my dad made into a pick-up. *(Bamonte photo)*

MY MOTHER ASKED ME TO SPEAK HERE TODAY; I WANT TO BE OBEDIENT, AND I WANT TO HONOR THAT REQUEST.

I STRUGGLED / WHAT KIND OF MAN WAS MY FATHER / MY THOUGHTS / HOW DO YOU PRESENT A LIFE OF 87 YEARS IN 10 TEN MINUTES

HE WAS JUST A SIMPLE MAN;

1. OUR FATHER TAUGHT MY BROTHER AND I IT'S OK TO LAUGH AT OURSELVES

2. OUR FATHER TAUGHT US NOT TO COMPLAIN

3. OUR FATHER TAUGHT US NOT TO BE JEALOUS

4. OUR FATHER TAUGHT US TO PUT GOD FIRST

5. OUR FATHER TAUGHT US TO TELL THE TRUTH

6. OUR FATHER TAUGHT US TO DEFEND THOSE WHO COULD NOT DEFEND THEMSELVES

7. OUR FATHER TAUGHT US NOT TO BLAME OTHERS

8. MY FATHER TAUGHT US TO ACCEPT RESPONSIBILITY

WE NEED NOT ENLARGE OR IDOLIZE MY FATHER IN DEATH, BUT JUST RECOGNIZE HIM FOR WHAT HE WAS, A SIMPLE MAN.

YOU SEE
I TOO WANT TO EMBRACE EMPATHY
I TOO WANT TO STRIVE IN HUMILITY
I TOO WANT TO BE CARRIED BY HUMBLENESS
I TOO WANT TO POSSESS SELF KNOWLEDGE
MY FATHER WAS ALL OF THESE
I TOO WANT TO BE A SIMPLE MAN

MY HOPE AND PRAYER IS THAT I WILL SEE MY FATHER AGAIN SOMEDAY

THANK YOU DAD FOR EVERYTHING YOU TAUGHT US

I LOVED YOU DAD, AND I WILL MISS YOU DEARLY

My Reasons for Becoming a Policeman
by Tony Bamonte

During my career in law enforcement I learned there were two types of policemen. Those who wanted to be in a position to help other people who are being victimized, and unfortunately, those with self-serving motives, who wanted the power of the position.

As a small boy, my family life was filled with habitual violence. Often the police would respond to calls at our house. During those incidents they did their job, but it was with kindness, compassion, and understanding they showed toward my sister, brother, and me. I grew up respecting police and someday wanted to be one. Fortunately, I have found there are far more good people in law enforcement than bad. A bad cop hurts his entire department. Those who are unprofessional in their actions or demeanor should be recognized, disciplined or removed by their commanding staff.

Louis Bamonte, 1927, as a marine. He was also a great role model. *(Bamonte)*

ARGUMENT OVER WIFE MEANS JAIL, HOSPITAL

One man was jailed on an assault charge and a second was taken to Sacred Heart hospital yesterday after the first reported he met the second coming from visiting his ex-wife and beat him up.

Louis Bamonte, 43, E517 Heroy, was in jail on the assault charge after, detectives said, he beat Henry Malina, N603 Oak. First he used a piece of a shovel handle, then a sharp piece of ice, it was said.

The detectives recalled that Malina had been beaten up several months ago, supposedly by Bamonte's father and brother.

The above article appeared in the December 26, 1948, *Spokesman-Review*. This happened on Christmas morning, when both our mother and dad went to jail. We went to live with our grandparents. It is inaccurate in the statement, "Beaten up several months ago by Bamonte's dad and brother." He actually was beaten up by my mother's dad and her brother. Two months later my grandmother and grandfather got in a heated argument in the middle of which my grandfather died from a stroke. The police also responded to that. In a court battle that followed, my father was given custody of all three children.

On a regular basis, police officers encounter people when they're at their worst – drug addicts, alcoholics, gang members, thieves, spousal abusers, etc. A good policeman should sincerely want to make a difference, set a good example, and remember he or she is a public servant, with the sole purpose of protecting the community.

As policemen, we may be the only good influence some of these criminals will ever see, typically because of their circumstances and the atmosphere they grew up in.

Tony Bamonte, 1947, with black eyes from abuse which occurred when I told my dad about a man my mother was seeing. She waited until he had left before this happened. My dad took this photo for evidence. *(Bamonte, evidence photo)*

More Family photos

My son Louie on my police motor in 1972. *(Bamonte photo)*

My son Louie, center standing, when he took up wrestling in high school. *(Bamonte photo)*

Louis and Tony Bamonte loading tamarack poles with a gin-pole in 1958. The truck was a 1944 GMC army truck and the dozer, an International TD 6. At the time I was about 16 years old. *(Bamonte photo)*

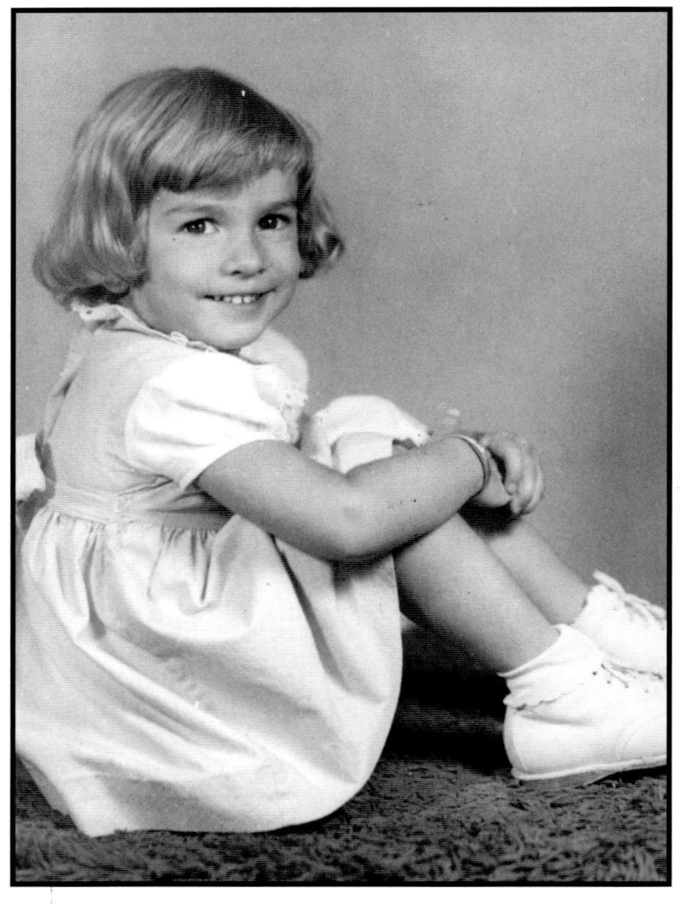

Suzanne Schaeffer (later Bamonte) at the age of three. This was also about the age she was when I first saw her. For about a year I delivered the *Spokane Daily Chronicle* to her families home. Suzanne had an older sister, Barbara, and two other younger siblings, Bette and Bill. Suzanne's mother, Mae, was involved in the girls scouts, along with many other worthy organizations. Her father was the plant manager for the Lehigh Portland Cement Company in Metaline Falls.

During the Christmas and Thanksgiving holidays Suzanne's mother would always make enough to share with some of the more unfortunate people in our community, having her girls deliver holiday meals to them.

In 1993, following a divorce from my wife of 22 years, I had an occasion to run into Suzanne again in Metaline Falls. We had both been attending a birthday party for a mutual friend. As I was leaving I noticed a car that was broke down a few block's away and stopped to help. It was Suzanne and her family. I gave them all a ride back to Spokane. *(Bamonte photo)*

During the time I was the elected sheriff of Pend Oreille County, as a side line and for exercise, I used to build log houses. I built a total of 11, almost one a year. Photo circa 1985. *(Bamonte photo)*

My brother Dale, circa 2015. He's a seasoned skilled union electrician. *(Bamonte photo)*

My sister, Star, who passed away in 2011. This was taken at her second wedding. *(Bamonte photo)*

Tony and Suzanne (Schaeffer) Bamonte, in 2005. Both grew up in Pend Oreille County.

Military Experience – the Days of the Draft (conscription) and Why I Joined the Army

The draft was one of the most significant crises that hung over the heads of every young man throughout the Vietnam War, going back to the Civil War. Every male in the United States was required to register for the draft within 30 days of his eighteenth birthday. Because of the draft, if a young man attempted to find work in any type of job that held a future, one of the first questions he was asked was if he had completed his military obligation. It seemed the companies offering the good jobs didn't want to hire young men and then lose them to the military. This was the reason many young men joined the army.

The small town I grew up in was a logging and mining community. It was easy to get a job in that line of work, but it was not stable and had no benefits. Consequently, after working at the Pend Oreille Mine for over a year, I enlisted in the army on the 20th of November in 1961. My choice for the army – was because it was an obligation of three years versus four years for all other branches of the military. If you were drafted you would only have to spend two years. The army was the only branch of the service that was drafting men. I felt it was worth that extra year to be able to get my military obligation out of the way. Although I had no idea of what the Vietnam War was at that time, there was also the strong concept of doing your duty for your country.

The Consequences of Avoiding the Draft

If you failed to register for the draft, the penalties were great. You would be considered a "draft dodger." Draft dodgers faced many direct consequences, which were enforced by the federal government: draftees who did not report as required were arrested and sent to a federal prison for a maximum of five years; draft dodgers lost their voting privileges while in prison; draft dodgers were also subject to a $10,000 fine.

In addition to this, there were many indirect consequences such as: loss of respect from fellow Americans; draft dodgers often left the United States in exile and were becoming illegal immigrants in the countries they were hiding; it was difficult for them to find a job; families were separated because they were exiled, they couldn't see their kids or family members; and most significant, they were often characterized as cowards and unpatriotic.

Ways Men Dodged the Draft

There were three ways most commonly used to avoid the draft during the time of the Vietnam War. 1. Flee the country to Canada or Mexico. 2. Enroll in college because students didn't have to register for the draft. The college exemption was often abused by many of the wealthy in this country who could afford to go to college. 3. Get a doctors note stating a reason you were unable to be in the military.

The End of the Draft

The draft changed the lives of many men in the United States. From 1940 to 1973, during both peacetime and periods of conflict, young men were drafted to fill vacancies in the United States Armed Forces that could not be filled through voluntary means. In 1973, the draft, which was highly unpopular, finally ended when it was repealed by an act of congress, changing to an all-volunteer military force.

Bamonte's Military Time 1961 to 1964

After enlisting in the army in 1961, I attended basic training at Fort Ord, California. From Fort Ord, I was sent to military police school at Fort Gordon Georgia. Toward the end of my training at Fort Gordon, I applied for Airborne. However, I was sent to Lackland Air Force Base to attend K-9 school. While there I was given the most beautiful German Shepherd dog, who would be with me for almost a year and a half. Upon completion of that school, my dog and I were assigned to Monroe, Michigan, to an underground Nike /Hercules Missile base.

I was at Monroe for approximately eight months, including the time of the Cuban Missile Crisis from

October 14, 1962 to October 28, 1962. Following the missile crisis I was transferred to another Nike/Hercules site, located on Belle Isle Park in Detroit. In my opinion Belle Isle Park is Detroit's nicest park, being located in the most serene, peaceful, and scenic parts of the island, with trails along the river and the Blue Heron Lagoon. During my time in the army, Belle Isle was the best location where I was ever stationed and the best duty, being with my dog.

I had been at Belle Isle about six months when I was promoted to corporal. Within weeks of that promotion I received orders for duty in Vietnam. The directions on these orders were for men with the rank of corporal. I was forced to give up my dog. All of the training and bonding was never considered. Fifty-four years later, a feeling of sadness washes over me when I think of the bond I had with him. I was told he would be assigned to another handler, but I miss him greatly to this day.

Vietnam

My arrival in South Vietnam was in July of 1962. I was assigned to the 560th military police company at Tan Son Nhut Air Base. Within weeks I was transferred to downtown Saigon, where I was assigned to MAC-V (Military Assistance Command Vietnam). MAC-V was the main headquarters for the brass responsible for the military actions being taken in South Vietnam. During my time working at MAC-V and living in Saigon, my residence was in a six-story hotel in the downtown area.

The MAC-V compound consisted of two buildings, one of which was used by the majority of the officers involved in the conflict and the other housed the offices of Ambassador Henry Cabot Lodge, Robert McNamera, Secretary of Defense from 1961 to 1968, and Commander of Forces in Vietnam, General Paul Harkins. Harkins was later replaced by General William Westmoreland, who had been his deputy.

My assignment at that location was at the entrance to both buildings. My job was to greet each officer with a salute and greeting. A number of things stuck out at that time: 1) The higher ranking officers, from colonels to generals, were far more polite and gracious than those of lower rank. 2) Henry Cabot Lodge, the ambassador to South Vietnam always rode in a solid black 1961 "Checker" automobile driven by his chauffeur. 3) Whenever the president of South Vietnam went anywhere he traveled in a motorcade made up of at least ten American-made, high-end, vehicles. During my time in Saigon I observed his motorcade on at least five different occasions. 4) The presidential palace for President Diem was far more opulent than that of the presidents of the United States. 5) Much of the life-style for the presidential family during the Vietnam War came from the American taxpayers military aid to South Vietnam. President Diem, and his family members, were highly inefficient and corrupt. Corruption abounded in all forms, especially with the large amounts of money the United States poured into the country in the form of military aid. President Diem's family was the biggest practitioner of nepotism. His close relatives filled the top ambassadorial, cabinet, and civil service posts.

By 1963, the number of US military advisors had grown to over 16,000. Also, by that time President Diem was perceived by the United States as an impediment to the accomplishment of its goals in Southeast Asia. Henry Cabot Lodge, the ambassador to South Vietnam, would play a part in his downfall.

Planing the Military Coup of President Ngo Dinh Nhu and the Very Small Part I Played

In early October, 21 days prior to the overthrow of President Diem, on November 2, 1963, I received an unexpected assignment. Along with five other men, all military police, we were moved to a house that was five houses away from the residence of Henry Cabot Lodge. Our assignment was to become familiar with the house that Lodge was living in, and when and if the time came, we were to protect and evacuate him if necessary.

During his time in Saigon, Lodge and his wife lived in one of the nicer areas of the city. It was located in a quiet residential section in a two-story house with a balcony. The commanding officer of our group of six had made arrangements for our meals to be furnished in the home, walking distance away, with an upscale Vietnamese family. This proved to be a good experience. At the time, we didn't know what we were going to protect him from. However, our job was to enable him and his wife to be evacuated from the roof of his home in Saigon. Our group of six men was there to provide cover for him. A helicopter would quickly swoop down and pick up him and his wife, delivering them to an undisclosed safe area. This was all preplanned, in case the coup went the wrong way and his life would have been in danger. Everything went as planned, and although ready, we were never needed.

Within days of the military coup and the killing of Diem and his brother, my detail came to an end, and I went back to my hotel in Saigon. At this point in my Vietnam tour I volunteered for duty as a helicopter door gunner.

Events Leading to the Coup

Diem's increasingly dictatorial rule succeeded in alienating the majority of the South Vietnamese people. Also, in the summer of 1963, Diem's brother, Ngo Dinh Nhu, had raided the Buddhist pagodas of South Vietnam, claiming that they had harbored the Communists that were creating the political instability. The result was massive protests on the streets of Saigon, causing a number of Buddhist monks to commit suicide in public by pouring gas on themselves and lighting it. Numerous pictures of monks engulfed in flames made world headlines. During the time I was in Saigon, on one occasion, I was a witness to one of these self-immolations by a monk. It was a sad and helpless feeling to watch.

The Actual Coup and the Killing of President Diem

During the summer of 1963, American officials decided the time had come for the Diem regime to be overthrown. Ambassador Lodge was to play an important role. When the time came, and with Washington's approval, the coup took place. Diem and his brother were captured and killed. Coincidentally, three weeks later, President Kennedy was assassinated on the streets of Dallas.

At first, the United States publicly disclaimed any knowledge of or participation in the planning of the coup that overthrew Diem. It was later learned that American officials had met with the generals who organized the plot and gave them encouragement to go through with their plans.

My Next Assignment
Helicopter Door Gunner Duty

Shortly after the military coup in South Vietnam, I volunteered to be a helicopter door gunner. I was assigned to the Utility Tactical Transport Helicopter Company (UTT) at Tan Son Nhut Air Base, which was an Air Force facility. The UTT was the first armed helicopter company of any army in the world of combat.

The UTT was originally organized in Okinawa and sent to Vietnam in September of 1962. However, the company's designation changed three times and was given a different name each of those times.

I served as a door gunner for most of February until sometime in March 1964, when it was my time to return to the United States.

My training as a helicopter door gunner lasted for three days. That training consisted mostly of first aid and how to administer morphine. During the first few weeks as a door gunner, I flew with various pilots and copilots. I was soon assigned as the door gunner for Major Patrick Delavan, who was the commanding officer of the UTT. I quickly learned that Delavan, by all accounts, was a legend as far as helicopter pilots go. He was fearless, to the point of seeming like he had a death wish. I later learned that none of the other door gunners wanted to fly with him because of that reputation. His helicopter was named Sabre 6, a number that was proudly passed on to other commanders.

An article about Delavan that appeared in the *New York Times* on June 12, 1964, stated: "Major in

South Vietnam Gets 7th Purple Heart." It went on to say: "... His eyelids were wounded when a rocket on his helicopter exploded after being hit..."

I was with him when that happened. I was his door gunner. As I recall the incident, the rocket did explode. A bullet hit the fuel propellant igniting the fuel, which in turn blew up. This caused severe burns to the crew chief and Major Delavan, who was directly in front of the crew chief. It also caused minor injuries to others in the aircraft.

In 1964, the United States was still in the role of military advisors. As such, we were required to have a South Vietnamese soldier accompany us on our missions. Also, whenever the media wanted to fly with us they would go with the flight commander. That particular day our helicopter had six people aboard. Major Patrick Delavan in the left front pilot's seat, to his right the copilot, behind him from left to right was the crew chief, a TV cameraman from one of the national news networks, a Vietnamese observer (referred to as an ARVIN), and then myself. When the rocket exploded it flared up discoloring the left side and left front plexiglas windows of the helicopter. The person injured the most was the crew chief, as the rocket that flared up, was just to his left and about three feet from him. If the rocket itself had exploded, as stated in the *New York Times* article, it would be highly unlikely that anyone would have survived.

Whenever we flew, both the left and right sliding back doors would be open. The crew chief almost died as result of this, and as far as I could remember, was in the hospital for several months. The next people to suffer injuries were Major Delavan, whose window was open at the time, and the news reporter, who was next to the crew chief was burned on the neck. The ARVIN was burned slightly on his left arm and hand. The copilot and I suffered slight injuries as we were the farthest away from the explosion.

What stands out most about that incident was when we were finally able to land, everybody exited the aircraft and we removed our flight helmets. When the TV cameraman took his off he looked quite different than he did prior to this incident. In 1964, the Beatles style haircuts were in vogue. That was the type of haircut the news man was wearing. As soon as he removed his helmet it was obvious his hair had been trimmed up a bit because of the fiery blast. That was the one bit of humor attached to that incident.

General Joseph Stilwell, Jr.
and the UTT Helicopter Company

General Stilwell served as commander of US Army Support Group, Vietnam (renamed US Army Support Command, Vietnam from 1 March 1964) from 26 August 1962 until 30 June 1964. During my time as a door gunner I came to know General Stilwell. Prior to my going home he asked if I would be his driver. If I had taken that assignment it would have been an immediate promotion to sergeant. However, I would have had to extend my stay in Vietnam for another year by reenlisting in the military. General Stilwell was lost at sea on 25 July 1966, when flying a C-47 to Hawaii with a longtime friend and pilot. The C-47 was to continue on to Thailand.

Total helicopter pilots killed in the Vietnam War was 2,202. Total non-pilot crew members was 2704. Based on a database from the Pentagon, an estimated 40,000 helicopter pilots served in the Vietnam War.

Tony Bamonte standing next to his assigned helicopter, Sabre 6, sometime in April of 1964. *(Bamonte photo)*

Thank you to the following businesses for their financial support of this book project:

Index